BRAIN RULES FOR WORK

brin
rules
for

work

The Science of Thinking Smarter
in the Office and at Home

Pear
Press

Pear Press
P.O. Box 70525
Seattle, WA 98127-0525
USA

This book may be purchased for educational, business,
or sales promotional use. For information, please visit
www.pearpress.com.

FIRST EDITION

Edited by Erik Evenson
Designed by Greg Pearson

Library of Congress Cataloging-in-Publication Data
is available upon request.

ISBN 978-1-7323803-8-7
10 9 8 7 6 5 4 3 2 1

Printed in the United States of America

To my dear friend, Bruce Hosford,
one of the warmest, kindest souls I know.

contents

10 brain rules for work

1.

Teams are more productive, but only if you have the right people.

2.

*Your workday might look and feel a little different than before.
Plan accordingly.*

3.

*The brain developed in the great outdoors.
The organ still thinks it lives there.*

4.

Failure should be an option—as long as you learn from it.

5.

*Leaders need a whole lot of empathy and a little
willingness to be tough.*

6.

Power is like fire. It can cook your food or burn your house down.

7.

*Capture your audience's emotion, and you will have their attention
(at least for ten minutes).*

8.

*Conflicts can be resolved by changing your thought life.
It helps to have a pencil.*

9.

*You don't have a "work brain" and a "home brain."
You have a single brain functioning in two places.*

10.

Change won't happen out of determination and patience alone.

introduction

I ONCE STARTED A LECTURE to a group of business students by asking them this question: "Why do we have five-fingered gloves?"

I waited to see if anyone would take a stab at the answer. When all I got were a few laughs and confused looks, I answered my own question: "Humans make five-fingered gloves because humans have five-fingered hands!" This prompted more laughter and, I'm sure, a little more confusion. After all, they were there to listen to a neuro-scientist talk about the business world, *their* business world in a few short years. What would gloves and fingers have to do with the work-place or their brains or both?

"Well, *duh*," I intoned. "Your brain has the cognitive equivalent of five fingers. The organ is shaped to respond with great productivity to certain environments and to other environments, with no produc-tivity at all." I reasoned that ergonomics applies to the mind just as much as to the hand. "If you're designing a workplace and want to optimize output, you better keep the cognitive shape of the brain in mind," I warned.

1

I went on to explain that the typical place of business is *not* designed with the cognitive equivalent of a five-fingered glove. And so I invited my temporary class to a thought experiment: What if the workplace *were* tailored to the brain, the way a glove is tailored to a hand? What would organizations look like if the business of profit-making took the business of brain function seriously? How would we design management structures? What would physical workspaces look like? What environment would best aid creativity, productivity, and the simple ability to just get things done?

The goal of this book is to answer questions just like these. We're going to explore how the application of the behavioral and cognitive neurosciences can improve your productivity in the workplace. The information is relevant whether you're working from the corner office at headquarters or from your closet office at home. Call it an exercise in cognitive ergonomics.

This won't be your typical book about work, however. Almost every concept here was braided by the deft hands of Charles Darwin. We're going to use his evolutionary ideas to outline the book's central challenge: how to work with a brain that operates in the twenty-first century but still thinks it lives in the ancient Serengeti. We're going to explore how this jiggly three pounds of incredible problem-solving genius, finely tuned to spear mastodons and pick berries, learns instead to run staff meetings and read spreadsheets.

Sometimes the organ conforms only reluctantly. After all, our brains haven't been exercising long enough in the civilization gym to completely break with the shackles of the Pleistocene, the prehistoric era in which the organ evolved inside the skulls of the first modern humans. Sometimes the brain conforms willingly to modern life, especially if we understand enough about its inner mechanisms to work *with* rather than *against* its natural tendencies. In a nutshell, we're going to explore how the science side of behavior informs the business side of behaving.

This effort is divided into ten brain rules. These rules are things we know about the brain based on peer-reviewed science. You can apply each rule to the corresponding area of your work life. Some rules deal with specialized areas of business, such as hiring practices and presentations. Others deal with more general areas of interest, from workspace design to getting along with others. We'll find out why you're so tired after Zoom meetings. We'll examine what you can do to your office, whether at home or at your work building, to make you more productive (hint: add plants). We'll learn why people become more interested in sex after they've been promoted. We'll explore the cognitive neuroscience behind creativity and teamwork, and discover the most effective ways to kill your PowerPoints. We'll end with explaining why good, old-fashioned change is so hard for good, old-fashioned people. With this knowledge, we will discover how to work smarter—designing our five-fingered glove, stitch by stitch.

The brain is amazing

Let's begin with some background information, starting with a few words about me, your glove outfitter.

I'm what's called a developmental molecular biologist, with special research interests in the genetics of psychiatric disorders. Those interests have manifested in two ways in my career—on the scientific side, as an affiliate professor at the University of Washington (department of bioengineering), and on the business side, as an analytical consultant, mainly advising for-profit companies in the private sector. The latter side of that experience is why I was asked to lecture to that group of business majors I mentioned.

I have been interested in taking cues from brain science and applying those cues to aspects of our lives for my entire career. In fact, I've written three books doing just that: *Brain Rules*, *Brain Rules for Baby*, and *Brain Rules for Aging Well*. I have never ceased to be amazed by what the brain can teach us. To illustrate this fascination,

I invariably start with a case study, whether giving a lecture or writing a book. These pages are no exception.

Let's talk about an unremarkable fellow with a truly remarkable concussion. Jason Padgett was a below-average student and college dropout, interested primarily in his biceps and his mullet. He hated math, loved *girls*—his word—and pretty much lived to party. At one of those parties, Jason was brutally assaulted and knocked unconscious. He woke up in the ER with a severe concussion. The doctors injected him with a massive painkiller, then sent him home. He would never be the same again.

When Jason awoke, he started seeing outlines of people and then, weirdly, over the course of days, began drawing extraordinarily detailed mathematical shapes. One day, during his convalescence, he was sketching these figures at a mall. A man approached him, looked at his work, and struck up a conversation. "Hello, I'm a physicist," the man said. "What are you working on there?" The man then said something that changed Jason's life: "Looks like you're trying to talk about space-time and the discrete structure of the universe."

Jason was stunned. The stranger grinned. "Have you ever thought about taking a math class?" he asked.

Jason eventually took the physicist up on his suggestion and discovered something both amazing and funny: Jason the party animal had become Jason the mathematical genius. His quantitative superpower was the ability to draw mathematical fractals, which quickly developed into a wide variety of math skills. Researchers from Finland studied Jason's brain and discovered that his injury had given him an all-access pass to specific regions that previously wouldn't let him get past pre-algebra. It was a mixed blessing, however. He also acquired obsessive-compulsive disorder and, for a few years, became a hermit.

Jason is a rare individual, diagnosed with Acquired Savant Syndrome, one of about forty such individuals described in the research literature. Math proficiency isn't the only acquired talent detailed in the literature. Other subjects with this syndrome have shown sudden changes in painting ability, writing proficiency, and mechanical aptitude. We have no idea how this shift happens. Padgett believes we all have certain hidden cognitive superpowers, if only we could find access to them.

That may be a bit of a stretch, but the possibility is intriguing and just one of the many reasons I'm so gobsmacked by the brain I haven't had a boring day in years.

(By the way, don't try Jason's model at home. Most people with injuries as severe as his don't wake up as Albert Einstein. Sometimes they don't wake up at all).

Energy hog

Understanding how researchers look at people like Jason requires some rudimentary understanding of how the brain works. Whether geniuses or not, we are all saddled with an unexpected tendency that borders on being annoying: Our brains are really into energy conservation. They function like a parent continuously nagging us to turn off the lights when we leave a room. The organ monitors how much energy your body is consuming, how much it's expending, and what needs to happen to fill up the tank. This accounting preoccupies so much of the brain's working life, some scientists are convinced that energy conservation is its *main* function. Here's how researcher Lisa Feldman Barrett puts it:

> *Every action you take (or don't take) is an economic choice—your brain is guessing when to spend resources and when to save them.*

The brain has profound reasons to be concerned about resources. It's an energy hog, acting like a three-pound SUV. It accounts for only 2% of your body weight, but the brain sucks up 20% of the available fuel.

That might sound like a lot, but 20% is barely enough to keep it functioning. The brain has too much to do (something to keep in mind about your audience if you're working on a presentation). It tries solving job overload by continuously scanning for shortcuts. For example, it reduces what it pays attention to, something best seen in visual processing. The eye initially presents to the brain the torrential equivalent of 10 billion bits of information per second, but then the brain's energy editors go to work. By the time the information reaches the back of the brain (the areas where you'll actually begin seeing something), that rate has been whittled to a paltry 10,000 bits per second.

The brain is so concerned with energy resources, it continuously livestreams to itself forecasts of how much energy will be needed for survival at any given moment. But it doesn't just forecast gas-tank information. Its predictive ability bleeds into many other areas, from predicting people's intentions to figuring out the best way to lead them—something that might be useful to know for those interested in becoming business managers or executives.

Sweet wattage

Exactly what type of energy resource does the brain consume? And what does it use that energy for?

The answer to the first question is familiar to anybody with a sweet tooth. The brain mostly consumes sugar (glucose), more than a quarter pound daily. The answer to the second question involves one word: *electricity*. The brain converts the sugar into electrical energy to perform most of its tasks, including communicating information from one brain region to another.

You can listen in on this electrical chatter by simply taping a few electrodes to your scalp. There's quite a bit to listen to, even when you think your brain's resting. It must keep many vital things going, after all: your heartbeat, for example, and your breathing—both of which require energy.

How much energy does it need? Stanford scientists estimate that a robot capable of accomplishing all the tasks a typical brain performs

6

at rest would require 10 megawatts of power—the typical output of a small dam. When the brain executes those tasks, it uses only 12 watts to do it—about enough for a small light bulb. No wonder the organ is so preoccupied with its energy supply!

How did our brains become both fuel-hogs and fuel-efficient? The answer comes from understanding a bit about our evolutionary past, a history we'll be revisiting in nearly every chapter of this book.

We'll discover we didn't start out with a powerful 12-watt brain. We started out with a much smaller version, one virtually indistinguishable from the brains of primates, descendants of which we can still see in the jungles of Central Africa today.

We'll also discover that, for reasons lost to antiquity, we began diverging developmentally from our simian sisters and brothers about 6–9 million years ago. We discarded our habit of walking on all fours, selecting instead the far more perilous bipedal design, which required us to continually shift our body weight back and forth onto constantly moving feet—a potentially hazardous development: we became top-heavy. Our incredibly important and incredibly fragile brains encased in our skulls (which account for a whopping 8% of our body weight) were now the body parts farthest away from the ground. Maintaining balance became a critical survival issue. Some researchers believe this shift produced a whole suite of demands on brain function, pushing us along the road to becoming earth's cognitive valedictorian. Our brains got bigger, more complicated, and more in need of fuel.

There is much that is controversial about this origin story and its timeline, as with just about everything else in hominid paleontology. In fact, the only thing scientists can agree on is that, for a while, standing didn't matter much. By the time we were 3 million years old, we'd only learned how to bang on things with partially chipped rocks. Yet things were about to change.

The first co-ops

A confluence of geological events more than 2 million years ago caused the earth's climate to transform dramatically, resulting in an overall cooling, and much of the humid African jungle home of hominids began drying out. Our once-steady climate became remarkably unstable. The aridification of Africa, resulting in an expanding Sahara, had begun, a process that continues to this day.

This aridification was potentially catastrophic for us. We'd spent most of our time enjoying a climate that was wet, humid, and relatively easy to survive. But now our situation had become more difficult. We could no longer just pluck food from trees, then wash everything down with a gulp from a nearby stream. We were forced to change from being creatures of forests to being creatures of grasslands. Our ancestors who survived the change from wet-wash to dry-cycle did so by becoming wandering hunter-gatherers, traipsing around a drier world—the African savannah. The requirements for that lifestyle changed nearly everything about us.

With the closing of our rainforest of a grocery store, we were compelled to walk longer and longer distances to find food and water. Such changes put new pressures on our developing, energy-guzzling brains. We really needed to (a) remember where we were, (b) decide where we were going, and (c) figure out how to get from where we were to where we were going. It's no accident that the same brain region involved in memory formation (the hippocampus) is also involved in helping us navigate flat surfaces.

Climatic change required us to learn how to navigate not only our physical surroundings but also our social relationships. The need to cooperate very quickly became a survival issue in the savannah. Why a survival issue? Compared with just about every other predator our size, we were (and are) a physically very weak species. Our canine teeth are so small and blunt that even chewing an overdone steak is challenging. Our fingernails (claws) don't do very well even against plastic packaging.

These deficits presented us with an evolutionary choice: we could get physically bigger, following, say, the elephants' size-upgrade plan. This would mean evolving to have a gigantic, dominating body column, which would have taken a gazillion years. Conversely, we could get smarter, shifting a few neural networks around, boosting something we were already starting to get good at: socially relating to each other. Such a shift would not take as much time as trying to become the size of an elephant but would have the same effect. It would create for us the concept of *ally*, effectively doubling our biomass without actually doubling our biomass.

Given that the estimated height of an average Pliocene hominid was 160 centimeters (about 63 inches), you can guess which path we chose.

Mammoth-sized cooperation

Cooperation turned out to be a useful design. It helped us to get otherwise impossible projects done, just as it does today. There are terrific examples of what groups of 63-inch-tall people can do if they learn to do it well. They become proficient in making death pits, for example.

A few miles north of Mexico City, a couple of these grim holes were found by a group of construction workers who were getting ready to dig out a landfill. The workers also found hundreds of mammoth bones, all concentrated into two pits, no creature showing signs of dying naturally. There were 14 individual mammoths altogether, along with ancient remains of camels and horses. These were hardly the only prehistoric killing pits ever discovered, but this was a particularly odd one: The animals had been slaughtered, butchered, flayed, and ritualized. One animal's bones were arranged in what the researchers called a "symbolic formation." The left shoulder of every mammoth was missing, leaving only right shoulders for investigators to ponder. All the mammoth heads had been turned upside down.

The researchers speculated that ancient hunters dug these exotic pits—which may have been filled with mud—then drove the animals into them, where they could be speared to death. At 6 feet deep and 82 feet long, the pits were certainly large enough for that. There's also evidence for a larger chain of pits besides these two, suggesting a vast killing field on an industrial scale.

The point? An adult mammoth stands 11 feet at the shoulder and weighs around 8 tons. No way a 5-foot, 3-inch human could take down even one of these animals solo, and remember there were 14 carcasses. For ancient hunter-gatherers to create a franchise of mammoth slaughterhouses, they would have needed to coordinate their behavior. Indeed, cooperativity was evident in virtually every physical feature of the Mexico City find, from digging the holes to carving up the meals to creating the rituals.

Aspects of this story are currently controversial and, clearly, will be the subject of more research. But one thing not in dispute is evidence of evolution's ability to transform a creature rising a mere 160 centimeters off the ground into the most colossal predator of the Stone Age.

Connections

Fast forward a few million years. We know today that the brain is one of the most powerful problem-solving tools evolution has ever fashioned. But how does it work? What are its quirks? Where is all the fuel going? And when we look into these incredible brains of ours, what do we find? Let's consider some basic brain biology.

It took many centuries to discover that this connectable kernel did anything important. It just sits there, after all, unlike hearts (which beat) or lungs (which bellow). As a result, most of the early research was simply a boring cartographic exercise. Early neuroanatomists cracked open the skull and named what they saw.

Many of the brain's structures were labeled after familiar objects from the non-brain world. For example, *cortex* means *bark*, probably because the brain's thin "skin" reminded some neuroanatomist of tree parts. *Thalamus* means *bed chamber*, possibly because somebody thought it looked like one (it doesn't, actually). *Amygdala* is the Greek word for *almond*, its shape reminiscent of the hard-shelled drupe. There's even a small pair of rounded structures called the *mammillary body*—so labeled, rumor has it, because they reminded the neuro-cartographer of his wife's breasts.

Early researchers believed that these regions were highly specialized, each with their own dedicated suite of jobs to do. They were partly right, but a modern understanding of how the brain works reveals a more nuanced and dynamic picture of brain structure and function. We now know the brain is not so much a collection of awkwardly labeled, unitasking regions, but rather hundreds of vast, dynamic, interconnected networks—the most intricate road map you'll ever see. There are clusters of nerve cells, many still corresponding to the old labels, which you can think of as being like cities. These cities are interlinked with miles of neural "roads." You have about 500,000 miles of these neural roads stuffed into your cranium, an object not much bigger than a cantaloupe. That's more than three times the total number of roads in the US National Highway System.

These networks aren't made of hardened asphalt, of course. They're made of squishy cells. Many different cell types exist in the brain, the most famous of which are called *neurons*. A typical neuron looks like a scared mop—an extended, furry head plopped onto the end of a long stick. You have about eighty-six billion of these oddly shaped cells stuffed into your head.

To form individual cables within the brain's networks, these mops are situated end-to-end, separated by tiny spaces called *synapses*. A typical neuron has several thousand of these synapses. These neural roads link together in astoundingly intricate formations. A handful of brain looks like the root ball of a rhododendron.

Wiring

Mapping such root balls is challenging, not that a lot of smart people aren't trying. Despite their efforts, which often burn through federal deficit-sized budgets, we don't yet have a complete, authoritative structural map of the human brain's circuitry. We call such maps *structural* connectomes. And the structures are not even the hardest part to chart. Even more difficult to chart are their functions, or the way specific circuits work together to provide some service. We call these *functional* connectomes. One reason these maps have proven so tough to make is a certain annoying generosity the brain possesses. That is, it offers many neural "employment opportunities" for the circuits lying in its interior.

Some circuits have job descriptions that are quite stable. They're hardwired into the brain and function similarly in anyone who's human. Consider, for example, two specific areas on the left side of your brain named Broca's and Wernicke's areas. These lefties are responsible for human speech. Damage the Broca's area in any human, and that person will lose the ability to produce speech (Broca's aphasia) but generally will still be capable of understanding spoken and written words. An injury in Wernicke's area (causing Wernicke's aphasia) does just the opposite, causing an inability to understand spoken and written speech but, astonishingly, does not impact ability to produce speech.

Such hardwired circuits are ridiculously specific, and not just for speech. Consider a man the research world calls RFS. Due to disease, RFS lost the ability to consciously comprehend numbers, and in a really weird way. If his brain detected a number, the image of the number got visually perturbed, flipping and then deteriorating into a messy visual blob. The deterioration never happened when his brain viewed letters, however. He could perceive, read, and write the alphabet just fine. His speech was great, too. The bottom line is that he suffered damage to a neural circuit that was specifically dedicated to processing numbers, separate from other visual inputs.

This is hyper-dedication on steroids. Yet such hardwiring is characteristic of only some brain circuits. Many are *not* laid down according to some universal human template. Some circuits display configuration patterns that are as specific to you as your fingerprint, which means every person's brain is wired differently from every other person's brain. So mapping the brain's structures and matching each to a function is achingly slow. Teasing out which circuits are common to everybody versus which circuits are common to no one has frustrated neuroscientists for decades.

Plasticity

Mapping is made all the more challenging when one considers that the brain is capable of rewiring itself on the fly. This might sound strange but is actually quite common. In fact, it's happening right now as you're reading this sentence. Whenever you learn something, the brain rewires itself. Whenever you process a new piece of information, physical connections between neurons change, sometimes by growing new connections, sometimes by changing preexisting electrical relationships. We call such rewiring *neural plasticity*. Eric Kandel won a Nobel Prize in part for discovering that, most of the time, the brain is hardwired to avoid being hardwired.

Know what this means? What you choose to expose yourself to profoundly affects how your brain will function. This can profoundly influence your relationship with stress and influence how much creativity you allow into your life – things we'll discuss in a bit.

The brain's ability to reorganize itself can be taken to ridiculous dimensions. Consider the case of a six-year-old boy who suffered from a severe form of epilepsy. To save his life, the surgeons had to remove half his brain (hemispherectomy). In this case, they removed the left side, the one carrying both Broca's and Wernicke's speech centers. You'd think with such catastrophic removal of hyper-dedicated neural tissue, the boy would not be able to talk or understand speech for the rest of his life.

That's exactly what did *not* happen. Within two years, the remaining right side of his brain had taken over many of the functions of the left side, including the ability to generate and understand human speech. The then-eight-year-old's verbal capacity had somehow been "miraculously" restored!

Does that mean the brain is so plastic, it can detect deficits, turn itself into a temporary neural workshop, then physically reconstruct itself? In the case of this little boy, yes. And he's not alone. There are many such published restoration accounts in the research literature, all of them baffling. Said Johns Hopkins neurologist John Freeman, who does this line of work:

> *The younger a person is when they undergo hemispherec-*
> *tomy, the less disability you have in talking. Where on the*
> *right side of the brain speech is transferred to and*
> *what it displaces is something nobody has*
> *really worked out.*

These are only some of the challenges researchers face as they attempt to create a comprehensive connectome map, a goal we are possibly still years away from achieving. Yet we are hardly clueless about how the brain works. Researchers in my field have chosen to specialize, deploying the scientific equivalent of a divide-and-conquer strategy. Exactly how that works—and how things are changing—is what we'll examine next.

Historically, we separated our investigative efforts into three distinct domains: The first domain was populated by researchers studying brains at the molecular level, researching how tiny snippets of DNA contribute to brain function. The second domain was populated by researchers studying brain function at the level of the cell— those tiny, frightened mops we discussed a few pages back. These cells could be examined at the individual-mop level or as groups of mops, meaning networks. The third domain was populated by people studying brain function at the behavioral level. This domain is the realm of experimental and social psychology. We'll talk about their efforts in virtually every chapter of this book.

The partitions between these molecular, cellular, and behavioral domains have blurred as the years have passed, thankfully, with many researchers actively pursuing questions in multiple domains. We even have an umbrella term for this blending, one we'll use throughout the book: *cognitive neurosciences*. This field of study is populated by scientists interested in linking biological process to behavior. The messiest research by far is the behavioral work, which deserves some special mention.

Skepticism and the grump factor

My scientific career often involves consulting with business professionals on issues related to human behavior. We usually end up discussing how to look at brain research with a healthy dose of skepticism. I'm a nice guy, but as a molecular biologist interested in psychiatric disorders, I can be pretty grumpy about what research says (and doesn't say) about the complexities of human conduct. There's lots of high-fructose nonsense out there, especially in the realm of self-help advice. One client called this skepticism the MGF, the Medina Grump Factor. It simply meant that facts I share are evidence-based, supported by peer-reviewed studies, often replicated many times. Just like most other scientists.

The same grumpy filter is true of the information in this book, but to keep it reader-friendly, I've chosen not to embed the references directly in the text. You can certainly see them for yourself. I encourage you to look up any of the studies mentioned within this book on the reference page at brainrules.net/references.

So what do I teach my business clients about applying brain research to their worlds? Realizing you can't build a career based solely on raising your middle finger at popular mythologies, I tell them to remember the following four issues.

Issue #1: *The field is still immature.*

We're still in the beginning stages of understanding even basic brain functions. We still don't know, after all these years, how your brain knows how to sign your name or how it remembers to pick up the kids at 3:00 p.m. It will be a long time before brain science can tell us what makes a great leader and what makes a great parking attendant.

Issue #2: *Many results are hard to reproduce.*

Human behavior is messy, and sometimes researching how it all works is just as untidy. Consider this ugly finding that rocked the behavioral research world a few years back: we're not always able to replicate certain important results in experimental psychology. University of Virginia researcher Brian Nosek formed something called the Reproducibility Project, an effort to reproduce specific famous behavioral findings. He and his colleagues discovered that only 50% of published experimental psychological results could be successfully (and independently) replicated.

This auditing is a good thing, of course, though it ran shock waves through the discipline. Many scientists painstakingly revisited old research findings to find the owies, then amended conclusions where warranted. This exercise was admittedly frustrating. We already knew precious little about brain function. Yet even some of the findings we thought stable—even canonical—had to be revisited.

Issue #3: *The origins of behavior are complex.*

You may have heard about an old debate concerning *nature versus nurture.* For years, there was a partisan conflict between these two words, one side thinking behavioral origins were primarily genetic (nature), the other side thinking the origins were primarily nongenetic (nurture).

Researchers have signed a truce these days, surrendering to the fact that nearly every human behavior has both nature *and* nurture components. Scientists laboring away in their prized molecular, cellular, and behavioral fiefdoms did more insightful research once they had this realization, opening their borders and often participating in multidisciplinary projects. I, too, tell clients that almost every behavior they can think of has both nature and nurture components. The mystery lies in understanding the percentage contributions.

Issue #4: *There's an inherent problem with crystal balls.*

One last concern involves an issue I've added only recently to my conversations with clients. The main text of this book was written in 2020–2021, spanning nearly the entire arc of the COVID-19 infection cycle. Watching the global business world stagger as if punched in the gut by this invisible adversary was heart-wrenching. Researchers from many fields are still doing damage assessment— and probably will be for years—examining the long-term effects of viral-mediated social and economic disruption. Because of the pandemic's recency, rigorous and solid evidence about these effects is currently exceedingly rare. I've thus warned clients about relying too heavily on people looking at behavioral crystal balls to predict the future of work beyond COVID-19.

If past is prologue, most people will get it wrong anyway. Perhaps no greater example of the perils of prognostication exists than trying to understand the so-called work-life balance, an issue we'll take up in chapter five. (Spoiler alert: some people think the virus has changed things forever, but I'm not so sure.) Sociologists will understand the impact eventually. So will you and I, but those details must be left to chapters printed in publications far younger than this book. If it helps, we're not going to predict the future in these pages anyway. We're going to reimagine it.

Taken together—even with a heaping helping of Medina Grump Factor—I absolutely believe cognitive neuroscience has much to say to the business world. The evidence-based suggestions in this book are well worth examining, even trying out. The practice will go a long way toward illustrating what business would look like if someone gave it a cognitive five-fingered glove.

teams

Brain Rule:
Teams are more productive,
but only if you have the right people.

I ORIGINALLY MEANT to open this chapter with a line from Scott Adams. He's the creator of *Dilbert*, that hapless, cartoon business-denizen of the newspaper comics page. In one of Adams' comic strips, Dilbert's boss meets with Dilbert and his team to discuss some of their accomplishments and failures. After announcing the team's less-than-stellar performance, Dilbert's boss says, "I only brought one teamwork award mug, so you'll have to take turns drinking from it."

Now I think a different opening is required. I'm going to start with a description of the animated short film *Bambi Meets Godzilla*.

Bambi Meets Godzilla starts with a lengthy opening-credits crawl. Bambi is contentedly munching on grass, pastoral music softly playing in the background. A minute into this idyllic scene, Godzilla's giant, scaly foot suddenly appears and squashes Bambi flat. This violence is followed by the words *The End*. The credit crawl quickly reappears, thanking Tokyo for its help in "obtaining Godzilla for this film." Everything fades to black.

Why start with Godzilla and not Dilbert? Because for those jobs still requiring in-person interactivity, an equally powerful and unexpected foot seemingly squashed flat that concept of work in 2020. That foot belonged to COVID-19.

Given this turbulence, can anyone say with authority what it *now* takes to make teams work effectively? Is anything in the cognitive neurosciences relevant to our discussion?

The happy answer is yes, at least regarding the behavioral sciences, and for a very specific reason: Darwin is stronger than COVID-19. The teamwork dynamics and social cooperativity that fueled our pre- and post-viral business offices were also observable forty thousand years ago. Back then, interactive cooperativity allowed humans to scratch two important Darwinian itches: the need for food and the need for protection. We could not survive without teamwork on the difficult plains of the Serengeti. And we still can't, virus be damned, even in the difficult boardrooms of any company greater than the size of, I don't know, more than two people.

In-person collaborative efforts were already becoming the norm in pre-pandemic commerce, from small mom-and-pop shops to behemoth multinationals. A 2016 study from the *Harvard Business Review* examined company behavioral habits and found that the hours "spent by managers and employees in collaborative activities has ballooned by 50% or more." The study also discovered that, for many jobs, 75% of daily activities involved interacting with other people.

Even science was affected. When I first started my research career, I occasionally encountered single-author papers. Now, they're practically an extinct species. In 1955, only about 18% of papers in social science were written by teams. By the year 2000, it was 52%. An ecology journal circa 1960 had about 60% of its papers written by single authors. In this past decade, the number has dropped to 4%.

The contemporary practice of this ancient teamwork idea didn't guarantee that every group effort would be better than every individual effort, however. We've all been involved in team projects where it would have been better if we could just do the darn thing ourselves— no buddies allowed. Statistically, however, teams are more productive, which is why their use was growing pre-pandemic and will be growing again as we crawl out from under the viral shadow.

What separates a good team from a bad one? Though no surefire recipe exists for every company, research is clear about what separates highly productive teams from less productive ones. We'll explain this research, ranging from behavior to biochemistry. We'll discover that creating effective teams is a fairly simple task as we slowly recover from our enforced isolation. Note, however, that I did not use the word *easy*.

To team or not to team

We should begin by asking a few questions: Just how effective are teams? Do they really make us more productive? Let's see what happens when we sidle up next to our coworkers at, of all places, the cafeteria table.

A study from Arizona State University showed that employees who ate their lunches at tables built for twelve rather than at tables built for four were individually more productive. Ben Waber from MIT's famed Media Lab speculated that this outcome was "thanks to more chance conversations and larger social networks." It seems that spontaneous interactions with coworkers encourage productivity. Waber found that companies that make it easy to bump into each other and interact, "with things like companywide lunch hours and the cafés Google is so fond of, can boost individual productivity by as much as 25%."

That's a pretty staggering number, but spontaneous interaction isn't exactly the same thing as team productivity. Luckily, there is a surprisingly large body of research that supports the idea that groups are better problem-solvers than individuals. They're more creative. They're better at spotting errors. They're more intelligent. Profitability rises when employees are encouraged to collaborate in groups. Employees seem to concur, according to one study. When asked what had the greatest impact on their company's ability to make money, 56% of respondents said it was collaboration. One reason COVID-19 presented such a business hazard is that the social isolation required to beat it put these numbers at risk.

Not everybody is as enthusiastic as MIT or Google about teams, however quantifiable those findings. One famous dissenter is J. Richard Markham, a research psychologist at Harvard, who's studied group interactions for a long time. He finds that most groups actually don't work well together. Infighting (competition for credit), asymmetric job distribution (some teammates do all the work), and confusion about aims (little agreement about what goals are to be accomplished) sandpaper away most of the benefits groups might otherwise provide. In an interview with the *Harvard Business Review*, Markham said, "I have no question that when you have a team, the possibility exists that it will generate magic ... But don't count on it. Research consistently shows that teams underperform, despite all the extra resources they have."

Markham doesn't entirely dismiss working in teams. In that same interview—perhaps inadvertently—Markham points to a way out. He says the main reason teams fail is a lack of trust between the members.

Luckily for us, we can measure the amount of trust there is in a team, and we have the ability both to detect the presence of bad teams and to measure the success of good ones. The types of metrics available to us are broad, ranging from the behavioral to the biochemical. We'll describe a tiny molecule, for example, that won a Nobel Prize for the guy who discovered it, something he accomplished mostly through a team effort.

The wisdom of Aristotle

When researchers looked into why we cooperate so well, relative to asocial species, they ran into what might be the friendliest molecule ever to percolate through the human brain. It's called *oxytocin*.

Oxytocin does many things for us, but one powerful behavioral feature is its ability to induce trust in people. In an experiment, subjects were given nasal spray filled with oxytocin. After they sniffed it, they were much more likely to trust strangers with their money. Researchers call this tendency "enhanced social learning." The pull of our social needs exerts a force over even our biochemistry!

The researcher who has championed much of the work exploring the relationship between feelings of trust and oxytocin also happened to garner a "sexiest man of the year" award. We scientists don't win many of these awards, which is why, when it happened, we perked up our ears. The Geek God, named "One of the 10 Sexiest

Geeks of 2005" was Paul Zak of Southern California (of course). When Zak isn't winning awards for his looks, he is also the world's authority on oxytocin and behavior.

Zak's work does much to reconcile MIT enthusiasts who think groups are problem-solving machines with grumpy Markham, who says things like, "No, they're not." Both would do well to examine one of Zak's most interesting oxy-related findings. He's discovered this relationship between oxytocin and interpersonal stress: stress hurts oxytocin production. Without oxytocin, engendering feelings of mutual trust becomes harder, which is why stress often hurts relationships. That finding is directly related to what makes some teams work well while others founder.

Cue Project Aristotle from Google, a research effort that put Zak's biochemistry findings to the test—and inadvertently also confirmed the reason for Markham's grumpy negativity.

Run out of Google's famed People Analytics division, Project Aristotle asked the company to do some navel-gazing. They looked internally to see what separated Google's malfunctioning groups from its most productive superstar teams. They found that the biggest difference was psychological safety, which turned out to be right in Zak's wheelhouse. Why? It involved trust.

An emotional climate that made "interpersonal risk-taking" feel safe for every member was the single greatest factor in Google's superstar teams. Other factors certainly mattered—ranging from being punctual to having shared beliefs about goals—but none of these came close to the importance of the trust each team member felt for one another.

Other researchers had been finding the same thing. Perhaps the most granular of these studies came from Anita Woolley, then at MIT. Just like the Project Aristotle team, Woolley was interested in what made smart and productive teams so smart and so productive. Was there a group intelligence that could be quantified, separate from the intelligences of individual members? Was there an emergent secret sauce that was present only when everybody got together? Was the whole greater than the sum of its parts? As you may know, Aristotle was famous for asking just this question.

I wonder if Google knew that.

The three elements of c-factor

To answer Aristotle's ancient question, Woolley and her colleagues examined the group behaviors of almost 700 people. She separated them into teams, then assigned them a series of tasks. Each task required a different set of collaborative skills, ranging from conjuring creative solutions to solving a thought-problem to planning a trip to the grocery store.

Sure enough, some teams worked really well together, but others, not so much. What made the successful groups so successful? The data appeared confusing at first. Some were managed by strong alpha-type leaders, while others had the authority more evenly distributed. Some teams had intelligent people who deliberately divided the solutions into bite-sized tasks. Others divvied up the workload by considering individual team members' superpowers before assigning jobs. Lots of variation. Not a lot of insight. There weren't any obvious commonalities that predicted success—until the researchers started looking at relational issues.

The one superpower all the successful groups had in common was how they behaved toward each other, how they treated one another relationally (with an interesting demographic twist we'll get to shortly). From this relational treatment sprang an Aristotelian group intelligence. And it was the degree of group intelligence that predicted how successful a team became. They named this group intelligence "c-factor," short for *collective factor*. The higher the c-factor in a group, the more successful the group was at their tasks, no matter how sophisticated or mundane. The difference wasn't trivial.

C-factor consists of three elements. You can think of it as a stool supported by three legs. In Woolley's study, all three elements needed to be present simultaneously to support the weight of the superstar scores. I'm guessing you'd like to know the components of c-factor. Without further adieu, here they are:

1. *Members of the group excel at being able to read each other's social cues.*

2. *Members of the group take turns in conversation.*

3. *The more women in the group, the higher the c-factor.*

What Borat has to do with teamwork

The first leg of the stool concerns something called Theory of Mind, a complex cognitive gadget that comes as close to mind-reading as neuroscience gets. It may be best illustrated with help from famed comedian Sacha Baron Cohen.

From Borat to Ali-G, Baron Cohen's comedic oeuvre often includes emotionally tone-deaf characters. Baron Cohen showcased one such character in interviews while he promoted his movie *The Dictator,* in which he literally played a dictator. He often gave the promotions in full character. One such interview involved legendary comic and TV talk-show host Jon Stewart. Dictator Baron Cohen started the interview by pulling a gold-plated pistol out of his waistband and laying the weapon on Stewart's desk. The audience gasped.

Things went casually downhill from there. The dictator discussed subjects ranging from jarring descriptions of his various sexual exploits to the loss of his dictator friends, Kim Jong-il and Muammar Gaddafi. Throughout the interview, Baron Cohen's dictator was blissfully unaware that his conversation was making the audience cringe. Stewart had a really hard time suppressing his laughter, perhaps because the stereotype seemed only too true.

What was Baron Cohen's dictator missing? Scientists would say he was missing Theory of Mind. Unlike Baron Cohen's character, people who have a strong Theory of Mind are good at detecting the emotional information in someone else's face. They also have the ability to take the perspective of others.

How do we know this? Though the ability to read facial information might seem different from the ability to shift perspectives, research reveals they both originate from Theory of Mind, a capacity to understand the psychological interiors of other people, their intentions, and their motivations. The core talent is discovering the rewards and punishments inside someone else's mind, thus developing a "theory" of their mind. To develop this theory, our own minds pick up on a variety of bodily cues, the most prominent being facial expressions. Extracting information from faces is such a big deal to the brain, it devotes an entire region (the fusiform gyrus) just to processing them.

Theory of Mind can be measured quantitatively, which is how researchers detect when changes occur. The psychometric measuring instrument is termed the RME, short for Reading the Mind in the Eyes. The RME is a test that shows you a series of faces of people experiencing emotions. Your job is to guess what the emotions are. There's a hitch, though. You can see only the person's eyes. People who have strong Theory of Mind do very well on this test. People with low Theory of Mind don't. The test is so robust, it's being used in some quarters to detect the presence of autism.

The author of the RME test should know. He's Simon Baron-Cohen, a brain scientist at Cambridge, and one of the world's authorities on autism. If that name sounds familiar, it should. He's the cousin of Sacha. I can't imagine what their family reunions must be like.

So, what does Theory of Mind and the RME have to do with c-factor? Woolley used the RME test as one of the metrics that tallied up what she called a "social sensitivity score," which is the first leg of the c-factor trifecta. Like coins from a slot machine, success tumbles out of groups with high social sensitivity scores.

Don't talk with your mouth full

The second leg on this stool is conversational turn-taking. To know what this is *not*, we need only turn to game night with my family.

When I was growing up, one of my family's favorite games to play was Pit, which simulated the open-outcry shouting matches of the old commodity-exchanges, before computers muted the brokers. In the game, you had to yell over everybody else, trading cards, interrupting, dominating, attempting a "corner" on the market. Our family's version got particularly zesty, and we often couldn't make out who was saying what.

One reason for the cacophony was that Pit creates a form of conversational turn-taking that is neither conversational nor turn-taking. Woolley found that teams functioning more like the game of Pit, where participants are jockeying for speaking time, were seldom productive, as my family can readily attest. She also found that groups with high c-factor did the opposite of open-outcry. No one person dominated the conversation when problem-solving. Everyone took turns deliberating. This metric was measurable in "air time" (speaking turns). Woolley wrote that "groups where a few people dominated the conversation were less collectively intelligent than those with a more equal distribution of conversational turn-taking."

Yep. Allowing certain individuals to constantly dominate the airwaves during a team meeting makes the group "less collectively intelligent." That's a ten-dollar term for *dumber*.

Another factor important to the concept of conversational turn-taking concerns interruptions, something Pit is very good at fostering. It's actually possible to measure interruptions using something called a *response offset*, which is the amount of time between when someone stops talking and another person begins responding. In normal conversation, response offsets are about half a second long. When people are interrupted, the response offset is zero.

Zero-offset responses are most easily observed in mixed-sex groups. This phenomenon was measured, remarkably, through an examination of transcripts from the US Supreme Court. Researchers found that female justices were interrupted by others 32% of the time. And they didn't return the favor: female justices interrupted others only 4% of the time.

Outside the courtroom, the same thing was observed. One experiment assayed interruptions in three-minute bursts. Men interrupted women twice when they conversed in this three-minute period. Men interrupted other men only once in the same time period. On average, men interrupt women about 33% more than they interrupt men.

Why does conversational turn-taking make such a difference? When people are allowed to speak, they have a chance to feel heard, safe, and as though their opinions matter. When conversational turn-taking is uneven—where one person dominates (or interruptions are frequent)—others may feel only approximations of these things. It dawns on this silent majority that their opinions may not matter as much as others' opinions, which is probably one reason trust is such a big factor in high-functioning groups. Remember, lack of trust is the reason groups fail. If nobody's dominating and nobody's interrupting, feelings of trust have chances to flourish. So does productivity. Not a commodity trader in sight.

Presence of women

The third leg of the c-factor stool may end up being the most controversial. Woolley found that increased c-factor scores were positively correlated with the presence of women. The more women in the group, the higher the c-factor became.

The reason? Well, that's the source of the controversy. Woolley says it's "because (consistent with previous research) women in our sample scored better on the social sensitivity measure than men."

The women Woolley tested in the group all had higher RME scores, what she calls "social sensitivity" (Theory of Mind).

The important line here is "consistent with previous research." She's referring to studies that demonstrate women tend to do better on RME tests. According to Oxford researcher Robin Dunbar, this edge is because women on average score higher than men in what are called second- and third-order Theory of Mind tasks.

Woolley may also be referencing findings that could easily be included in c-factor's second leg. Researchers have known for years that men's interactive styles in Western business cultures tend to be filled with social-dominance cues. This includes a willingness to give orders and using nonverbal cues such as chin-thrusts, directed eye contact (something researchers call *foveating*), and aggressive gesturing with hands and body.

Conversely, women aren't nearly as autocratic in their business behavior. They can make tough decisions with enthusiasm equal to that of their male counterparts, but they choose a more democratic strategy initially. The behavior is strikingly egalitarian, seeking the needs of the group first, looking for consensus where possible. They're more willing to display invitational safety cues (smiling, for example, as opposed to chin-thrusting), signaling that interpersonal interactions are a priority. Like I said, controversial.

Matters are made more complex when we ask how many women a group needs. The answer is: the more women, the higher the c-factor. The data are actually dose-dependent—but only to a point. That point is reached when the group is composed of only women. Then performance flattens. This finding is consistent with a large body of work demonstrating that diversity in team settings is a huge factor in team success.

C-factor may be only one component. That's a subject we will turn to later in the chapter. Right now, I suspect you're wondering what you can do to raise your team's c-factor where you work. It begins with a simple phrase that was first uttered not by Aristotle but by one of the other great Greek philosophers, Socrates: "Know thyself." Specifically, try to think back to who you were and how you behaved in kindergarten, when your brain was still tucked inside a rookie human being. So many of the devices you employ to get along with others today were developed long ago, for better and, sometimes, for worse.

Kindergarten

Most behaviors forecasting your business success were laid down when you were young. Neuroscientists know so much about them, we can even predict your future economic success based on your kindergarten behavior. It took researchers almost three decades to find this out.

In Canada, 3,000 preschoolers were monitored for their social interactions. Researchers examined "prosocial behaviors" (cooperativity, the ability to make/keep friends), "antisocial behaviors" (aggression, general opposition), and "focusing behaviors" (inattentiveness, hyperactivity). The researchers asked, "How will the kids turn out later in life?" then waited thirty years for their longitudinal answer. Many of the metrics concerned job and economic achievement. Did any behaviors predict success? failure?

The answer was a resounding yes, to both issues. The kids who were inattentive in kindergarten almost always had a lower income thirty years later. Those who were aggressive and oppositional not only earned less later in life but also were more likely to go to jail, suffer from substance abuse, or both.

The converse was also true. The more attentive the kids were in class, the higher their overall income became. The more prosocial skills they developed, the more likely they were to make friends. They also did better at school (which is related to one's ability to make friends, believe it or not), which means they were more likely to go to college and earn a higher income.

These findings are only correlative examples of a rather large mountain of data that we can summarize in a single sentence: social skills matter to personal prosperity. They also explain why change is hard. Social skills, once shaped, exert a powerful force over our behavioral trajectory.

Fortunately, there are evidence-based methods that offer opportunities to change course. For example, we know how to improve Theory of Mind abilities (as much a skill as a trait). We also know what you must do to quit dominating conversations and interrupting people, especially if you're a man used to cutting women off mid-conversation.

None of this is easy. All of it is doable. You may be shaped by your past, but happily, you no longer live there. The present is all you have. It's also all you need.

Time to get practical. Here's what you can do next Monday.

Narcissism

Luckily, there are things you can do to strengthen each of the three legs of the c-factor. We'll start with Theory of Mind.

In the preceding section, I noted that Theory of Mind is as much a learnable skill as it is an inborn trait. So how do we improve this delicate mind-reading ability? The key is to focus on somebody other than yourself.

Do scientists know how to make people less self-centered? Two sets of experiments say yes, they do. The first experiment involved maladaptive narcissists, the most self-centered people on the planet. The second involved book clubs. Yes, book clubs.

The experiment with narcissists came from the UK. Two groups of people had their narcissistic behavior evaluated, including assessments of how they reacted to troubling, empathy-inducing narratives. Both behavioral and physiological evaluations (autonomic responsiveness, such as heart rate) were taken. In typical nonnarcissistic populations, there's always an acceleration of nervous system responses when exposed to empathy-inducing stories. It's a terrific way to evaluate someone's reaction without relying on biased self-reporting.

The experiment began with narcissists in both groups hearing harrowing narratives, including a domestic-abuse survivor story and a severe relational break-up story. The first group, the controls, were then asked neutral questions such as "What did you watch on TV last night?" Immediately afterward, their brains and bodies were assessed. In the fashion of true narcissists, they were unmoved by the stories, just as they were by their television habits. Their physiological responses remained unchanged.

The second group was also asked questions, but these questions were devoted to penetrating the story they'd just heard. The subjects were challenged to imagine what the person might have felt when experiencing the trauma, and how they felt describing it. They were asked, "How would you feel if it had happened to you?" Anything to force-shift the focus onto the other person. Their behaviors and physiological responses were then assessed.

Sure enough, it worked. Empathy scores skyrocketed. Their cardiovascular reactions followed suit, experiencing a 67% increase in autonomic response over controls. The researchers "showed that, when instructed to take the perspective of a suffering target person, the deficit in empathy and HR (heart rate) associated with (maladaptive) narcissism is eliminated."

Yep, eliminated. Even highly insensitive populations can be moved. Surprisingly, it didn't take all that much to move them. Even brief instruction began changing their nervous systems.

These data underscore a double theme: One of the most toxic ingredients of a low-functioning team is self-centeredness. One of the most powerful antidotes is other-centeredness. Everything would be better if people got into the practice of regularly inhabiting someone else's world, then contemplating what it would be like to live there—not an easy thing to do on your own but not impossible.

Researchers know how to make people less self-referential, which means there is practical ore we can mine from these data. And this mining can be done without becoming someone's counselor or parent or other authority figure. Oddly enough, this point is illustrated by the fifth highest-grossing movie of 1984.

Waxing on and the power of literary fiction

The movie to which I'm referring is the original *Karate Kid*. The plot concerned an old, wizened-as-a-walnut karate master, Mr. Miyagi, teaching high schooler Daniel LaRusso how to perform martial arts. Mr. Miyagi began by asking Daniel to do repetitive, boring jobs around his house (like painting his fences, sanding his floors, and, memorably, waxing his car), practicing specific suites of motor skills. The idea—implausible now as it was then—was that when Daniel was doing these chores, he was actually learning a foundation of martial arts moves that helped groom him into a tournament fighter.

Believe it or not, Mr. Miyagi was relying on a process we call *far transfer,* where the practice of one skill inadvertently leads to proficiency in another. Though doing chores in order to learn karate is a little far-fetched, the idea of far transfer is not implausible. Which brings us to book clubs. The idea of book clubs helping a cognitive skill is an illustration of this transferring principle. By reading good books, you improve Theory of Mind.

Want proof? In a fascinating quintet of experiments, researchers in New York tested a group of people for Theory of Mind, then had them read literary fiction. As with the narcissists, the subjects were asked to penetrate the narrative, discussing characters and predicting how they might act under certain circumstances. This exercise forced them to engage with the text in nonsuperficial ways, such as one might experience in an invigorating book club. And as with the narcissists, changes in behavior were noted. In this case, Theory of Mind scores shot up about 13%.

This is a great example of far transfer. The research succeeded because practice in one domain yielded benefit in another. Researchers believe this transference worked because literary fiction simulates the real world of human relationships, giving people practice focusing on other people. (Some researchers call fiction a flight simulator for the heart.) It then becomes more reflexive to focus on people in the real world. Interestingly, these experiments worked only if the books were well written: they had to have won some literary prize. Popular fiction didn't provide a boost. Nonfiction didn't either.

The implications? It borders on the ridiculous to say that teams might function better if they formed literary book clubs, then immersed themselves in the characters' lives, but that's exactly what the data suggest. They should form book clubs or form movie clubs or volunteer at food banks, then write about the people they observed, then read those missives to their colleagues. Since social sensitivity improves team productivity, the syllogism almost writes itself. If you want to improve productivity, have people regularly experience worlds other than their own.

Support, don't shift

Recall that the second leg of the c-factor stool involves conversational turn-taking. Everybody talks in turn, no one dominating. This behavior is admittedly rare, and for good reason: the brain is your natural enemy here. With rare exceptions, most people love hearing themselves talk—the ultimate public expression of you, your ideas, your self. It can be addicting. The brain gives you a squirt of pleasure-inducing dopamine every time you do. Just like cocaine does. And Twitter.

How do people stop dominating the conversation? Sociologist Charles Derber may know. He's studied and classified hundreds of conversations at home and at work, confirming quantitatively that people really do love talking about themselves. His work also shows a practical way out, which can be illustrated with the following hypothetical chat between two coworkers:

Person One: I'm so unhappy with Madison.

Person Two: I'm unhappy with Madison too. Know what she did to me this morning?

Did you catch what changed? Person Two immediately started talking about his own experience, though the chat wasn't about him. Derber terms this conversation a "shift response." It's because Person Two shifted the conversation to his own experience, disregarding Person One's experience.

Now consider this conversation:

Person One: I'm so unhappy with Madison.

Person Two: Why are you unhappy? What's been going on between you guys?

This is different. Person Two kept the focus on the colleague who started the conversation and did so in a supportive way. Derber calls this a "support response."

People engage in the self-occupying shift responses most of the time (about 60% in live conversation). Examine their social media conversations, and that number bumps up to 80%. Meetings are often public displays of this tendency, dopamine cheering from the neurological sidelines.

Where are you on the narcissism meter? To find out, simply monitor your own conversations during meetings, or ask someone to do it for you. Take an inventory, either formal (make a small chart if it helps) or informal (ask a coworker how long she thought you talked). Maybe have someone time you, assessing how much airtime you actually consume. And consider how much of it is focused on you. If 60% of the time you're doing shift responses, invert it so that 60% is filled with support responses. If 80% of your digital experience is about yourself, make that number 20%.

Much like the Theory of Mind material, continually moving the conversation away from your favorite subject—you—just might give other people a chance to contribute. And that, as the research clearly shows, gives everybody a shot at making everybody look good, thereby strengthening the bolts of c-factor's second leg.

More women

You might remember that the third leg of the c-factor stool was sex-based. The more women in a team, the higher the productivity. The effect was reliably dose-dependent, like medicine. The practical suggestion here is, thus, as obvious as a glass ceiling: Hire more women. Promote women into positions where they can actually make a difference.

While that might sound controversial, these data were not discovered in isolation. Nor were they discovered recently, nor in just North America. The Organization for Economic Cooperation and Development (OECD) noticed a sex-based effect in developing countries more than a decade ago (2010). Women, upon receiving capital, invested more money into their families and surrounding communities than men did, which resulted in greater local wealth for everyone. If women's land ownership was equal to that of their male counterparts, crop yields increased 10%.

Even in their younger years, women lift productivity. The OECD found that if a country allowed at least 10% of their girls to become educated, overall GDP rose by about 3%. Women, when taking a long-term place at the economic table, boosted the financial potential of entire nations.

Researchers noticed similar effects closer to home. Fortune 500 companies with a balanced composition of men and women on their boards made more money than those with an unbalanced composition, and the effect wasn't trivial. Compared with the unbalanced environs, the companies with the balanced boards had an average 66% boost in return on invested capital, a 53% increase in return equity, and a 43% uptick in return on sales. These balanced boards also got into less trouble with the SEC because they practiced less risky behavior when trying to reduce the tax load of a corporation.

Why does all this productivity occur? Nobody really knows. People tend to use data like these to load up their guns in the culture wars, which is almost pointless with data this clear. Woolley believes the productivity is tied in with the first leg of the c-factor: since women tend to score better than men on social-sensitivity tests (like RME), and higher social sensitivity increases productivity, it logically follows that more women equals a more productive group.

These data no longer have to shelter in place. If you want more productivity, hire as many women as you can.

Drawbacks

Though groups usually make better decisions than individuals, high-scoring c-factor teams best of all, collaboration is not a universal win-win. The downsides, which have also been studied by researchers, can be quite devastating to organizations. In what must sound like a contradiction, the solution is more teamwork.

One of the most widely characterized shortcomings of working in teams is *groupthink*. Behaviorists define the term as "the tendency of group members to suspend critical thinking because they are striving to seek agreement."

Though it sounds like something George Orwell might've written, the term was originally coined by Yale psychologist Irving Janis in the 1970s. It's gotten a lot of mileage since then, invoked to explain everything from doomed military invasions to the twin Space Shuttle disasters.

Janis discovered that groupthink thrived only under specific social conditions. The gateway drug was information constriction. Teams walling themselves off from outside inputs were the most vulnerable, and for an obnoxious reason: they often grossly overestimated their competence. Because they were quick to compare themselves with others, especially if they'd enjoyed a past history of success, they were also vulnerable to developing a we/they tribalist mentality. The first casualty? Dissent. It became easy to mislabel alternate outside influences as "other," a euphemism for *inferior* or *threatening*, or both.

Another soldier in the groupthinking battalion involves external pressure. Teams compelled to create solutions within a specific time frame ("I need it yesterday") are more vulnerable to groupthink. This pressure may be supplied by an authoritarian-type leader, which is another big, flashing red light. A dominating personality, especially in a leadership role, is yet another risk factor for groupthinking teams. The need to please that leader, especially if the leader welcomes adoration, may take precedence over critical thinking.

One other odd thing about groupthinkers: they often work really well together. They may even pride themselves on what the military calls "unit cohesion," deeply reinforced if they've had a history of accomplishment. But long term, that cohesion is a mixed blessing. Alternate ideas in these teams may temporarily disrupt groupthink, but *blame* is mistakenly assigned to the critical thinker as those disruptive ideas, whatever their merits, are drowned out by the word *disloyal*.

And therein lies our contradiction. Isn't unit cohesion also a characteristic of c-factor? Doesn't c-factor promote safety, for which members may be so grateful that unity dangerously begins outranking critical thinking? It's almost like high-functioning c-factor groups need another ingredient—some other behavioral governor—to avoid groupthink. Researchers, happily, know exactly what is missing. More importantly, they know how to supply it.

The power of not looking like you

What's missing is best illustrated by what has to be one of the oddest couples in American jurisprudence history: the late Supreme Court justices Ruth Bader Ginsburg and Antonin Scalia. They were blazingly bright and fiercely independent. And they were as far away from each other politically as french fries are from carrot sticks. Yet their diversity drove them together, generating a fascination, a deep respect—even love. They socialized together, went to operas together (actually had one written about them!), and became best friends. At Scalia's funeral, Ginsburg delivered the eulogy.

This willingness to embrace differences lies at the heart of what's missing in groups addicted to groupthink: diverse of points of view—diversity of mind, thought, and perhaps most importantly, social experience. Racial, economic, gender, religious, linguistic, and—I'd argue—even geographical diversity all contribute to the richness and competence. There's a great deal of empirical support showing that the more diverse the group becomes, the better that group functions. They're less likely to become intoxicated with each other too, the normal anesthetizing elixir of groupthink.

One hint of this positivity came years ago. A group of researchers from Columbia University and the University of Maryland were interested in understanding what caused market crashes, looking carefully at ethnically diverse markets and price bubbles. Though the research was complex (primarily dealing with asset overvaluation), their findings revealed something very important: ethnically diverse markets resulted in more accurate asset assessments because they avoided the blinding overconfidence that groupthink produces.

How much more accurate were the assessments? A whopping 58%. This accuracy translated directly into saving a tremendous amount of money, millions of dollars in fact, as the researchers who conducted the study demonstrated.

Fact-checking behaviors were a significant reason for the higher accuracy of the assessments. The more social diversity a group possessed, the less bias-based errors they generated and the less assumptions they made. The voices of dissent were willing to challenge assumptions, resulting in a greater accuracy of facts. Diverse groups were more creative too, measurable by the number of novel ideas proposed per unit of time when problem-solving.

They unsurprisingly provided more innovative solutions to problems than homogenous controls did, and they made better decisions too.

The results are robust enough that researchers have been able to propose a mechanism explaining why diversity works, which we'll go through together. To be honest, though, they could have saved some time, money, and a lot of lab work. They could have just eavesdropped on the conversations between two Supreme Court justices attending an opera, at the height of their powers, at the depth of their affection.

Mechanisms of diversity

I was born in Japan, the son of a career military man. Of the top ten things I remember about my first few years there, kite-flying reliably vies for first place. These delicate objects looked so beautiful from far away, like someone smearing tiny, bright smudges of paint onto the sky. I remained enraptured by kites in college, but as my scientific understanding grew, I admired them for a wholly different reason: kites won't fly without tension. It's the wind blowing against their paper-and-wood skeletons that creates the lift necessary to make them soar.

From sermons to self-help books, tension-needing kites are frequently used as metaphors to teach life lessons. I'm about to recruit them for a similar social service, though on wholly different grounds. The kite principle is the reason why diversity can make teams super-solvers.

Someone who studied diverse super-solvers closely was the late Katherine Phillips, formerly of Columbia Business School. She found several interesting things about group dynamics, including that when socially diverse teams meet initially, there's almost always tension. Communications tend to be short and terse, the participants still wary of their new and unfamiliar surroundings. Many feel emotional discomfort. They also commonly have an apparent lack of trust, heightened concern for respect, and less group cohesion.

Given this emotional *ickiness,* you'd think diverse groups would be more likely to fail. That's the opposite of what's been found empirically, another great example of why even obvious impressions need investigating.

Phillips did investigate, and the answer she and her colleagues found is right at home with kite physics. The initial tension in groups puts people "on their game." Because of perceived differences, group members are more likely to change how they view expectations for group success. Some believe it will take more effort to reach an agreed-to decision. Some believe they'll need to pay more attention to getting their facts straight, leaving bias at the door. A self-cleaning dynamic begins forming. Phillips herself said, "Simply adding social diversity to a group makes people believe that differences of perspective might exist among them, and that belief makes people change their behavior."

Does this finding challenge the notion that trust is the most critical factor to group productivity? Not at all, though it does qualify it. The best groups still trust each other, but not because they're tension-free. They instead use the tension to become successful—as the literature points out, often wildly so—and nothing contributes to group cohesion like success.

Phillips discovered there was bad news, good news, and unusual news. The bad news was that tension precedes the initial interactions among diverse teams. The good news was that, with time, such teams become the best problem-solvers in the world. The unusual news was that it takes the bad news to produce the good news. It really is like flying kites. If you want to soar, you'll need to start with the wind not at your back but in your face.

On the matter of size

Teams, as we have discovered, can be the best problem-crushing productivity factories in the world. But we haven't talked much about how big they should be to optimize all that crushing. Is there a one-size-fits-all number?

The truth is that nobody knows—or, better, nobody's sure. Team sizes in the sciences sometimes seem to have gone off the rails. For example, the number of authors credited with discovering the mass of the Higgs boson (the poorly named "God particle") was more than 5,000. In my field of genetics, it's now common to have papers with 1,000 authors. Single-author papers are still observable, but they're as rare as Michelin stars.

Do large teams do any good? Researchers decided to find out. One effort examined worldwide research and engineering efforts over a sixty-year period (1954–2014), focusing on a deceptively simple question: What size of team does the most productive science? Investigators analyzed 65 million projects, looking carefully for any size-related trendlines. They found two.

The first trendline concerned how "disruptive" and "inspiring" a given research effort was. *Disruptive* meant unique/unconventional/newly upending. *Inspiring* meant examining how many times other researchers cited the disruption to inaugurate (or supplement) their own efforts. The study revealed that truly disruptive research almost always occurred within teams of less than five people. It didn't matter what research discipline or even project type. *Small* was an equal-opportunity characteristic of disruption.

Yet *small* wasn't a universal positive. Diminutive groups may have created more disruptive science, but they weren't really good at further developing their ideas. For that, larger groups proved necessary. Behemoths excelled at extending and elaborating on preexisting disruptions. Goliaths tended not to *generate* new ideas but be wildly successful at *making them work*. (It might have taken a couple of scientists to posit the existence of a God particle, but it took more than 5,000 to find it.)

One interesting footnote concerned the fate of individual investigators. When researchers followed the migration of disruptive scientists who left small groups to join larger teams, they found that the disruptors became less disruptive. Creativity was better served if those investigators stayed in smaller teams.

The conclusion? Both group sizes are necessary, which suggests that project managers need to be smart at the team-assembly stage. They need to assess the type of problem they have, then form the crew size most capable of solving it. The data show that the probability of team success (or failure) is determined at the point of assembly, way before the team is unleashed to solve problems. Assembling formidable problem-solving machines means creating teams high in c-factor and fortified with diversity.

What does that suggest you do next Monday? You should do the following:

1. *Select people who either score high on RME tests or are willing to join a literary book club to improve their scores.*

2. *Select people willing to examine their conversational habits. They should transform their shift-response verbal habits into support-response ones. They should practice conversational hygiene—not interrupting people, for example. They should be willing to learn to listen better.*

3. *Select for diversity, which runs the gamut from gender to race and from the geographical to the geopolitical.*

4. *Select for size: Small teams better serve creativity. Large teams are better at transforming that creativity into production-sized practicality.*

Most of these evidence-based ideas have deep roots in our long evolutionary history, giving them stability and providing us with a reason for hope in turbulent times. Even the severest damage that pandemics might inflict on group-based interactions will only ever be short-term. We have needed to work in teams for thousands of years. We will still need to work in teams for thousands more.

TEAMS

Brain rule: *Teams are more productive, but only*
if you have the right people.

1. A psychologically safe environment where coworkers trust each other is the key ingredient to a team's productivity.
2. To produce a psychologically safe environment, your team needs three things:
    1. The ability to read social cues (strong Theory of Mind)
    2. Team members taking turns talking without interrupting each other
    3. More women

- To improve your sensitivity to social cues, join a fiction-book club, volunteer at a food bank, or participate in any activity where you can practice focusing on someone other than yourself.
- To avoid groupthink in your team, employ a diverse group of people in thought, race, religion, gender, economic status—and maybe even geography. The initial tension of the team due to differences will help the team become successful in the long-term as long as the team maintains a psychologically safe environment.
- Choose the size of your team based on the project you're trying to tackle. Small groups (five or less people) work better for projects where the goal is to create something new or disruptive. Larger groups work better for building on preexisting creations and disruptions.

the home office

Brain Rule
Your workday might look and feel a little different than before. Plan accordingly.

When CEO Cathy Merrill published an op-ed in the *Washington Post*, she inadvertently stepped on a rake—and has the scars to prove it.

The title of her article, whose subject lies at the center of this chapter, said it all: "As a CEO, I Worry about the Erosion of Office Culture with More Remote Work." She bemoaned the loss of personal office interactions in the socially isolating world of COVID-19. No more spontaneous three-minute interactions in the office hallway. No more in-person meetings of any kind. She feared that, by the time things settled back to pre-pandemic norms, employees might be too used to the freedom of only occasionally stopping by the office.

She ended her lament with a bombshell: employees who wanted to work from home were in danger of having their jobs converted to contract work, she wrote. Paid hourly, they would be stripped of healthcare and retirement benefits. She said that the biggest benefit to coming back to a real office was simple job security. "Remember, something every manager knows," she concluded. "The hardest people to let go are the ones you know."

The result was like lighting a match in a propane tank. Sensing a not-so-veiled threat, her employees exploded in rage. Colleagues at other institutions were shocked. Merrill's workers eventually put out a tweet reading, in part, "We are dismayed by Cathy Merrill's public threat to our livelihoods," then went on strike for a day. The public furor lasted much longer. Merrill responded by saying she felt misunderstood, that her focus was mostly about "preserving the cultures we built up in our offices."

The irony here doesn't escape me: Any misunderstanding might have been cleared up quickly if everyone had been at work that day.

She could have convened a meeting, let employees air their differences, restated her intentions, then took everyone out for drinks (the first round on her). Instead, Merrill's workers, sheltering in place, reacted with a remote, smoldering fury. Employees were left humiliated. Merrill was left bruised.

How do we do meetings as employees slowly get back to work in such a frustrating post-pandemic world? What are the pitfalls if we continue to work remotely, in full or in part? Can they be avoided should remote work become a more permanent fixture in the American workplace?

This chapter addresses these questions. We'll start with familiar ideas about one of the staple activities in the offices of the pre-virus world: meetings. We'll examine how these have changed as more and more of us have conducted meetings over our computers from home. We'll end with thoughts about how to maximize your productivity at the home office. As you'll see, the home office can be as viable a workplace as the traditional office if you keep a few important points in mind.

The way we were

In the pre-viral world, there were two self-flagellating impressions about holding meetings.

The first was that meetings sucked, and not just in the pejorative sense of that word. They actually sucked resources such as time, energy, and finances. They were useless too: about 90% of people daydream during meetings, and more than 70% use meetings to get other things done. The second was that, for all their hollering, business folks held lots of meetings anyway—up to 11 million per *day*, up to 15% of an organization's time, up to 23 hours of a busy manager's week. The time spent from all these meetings could be costly too, with estimates crawling up to more than $37 billion per year.

This self-torture has produced a cottage industry devoted to teaching people how to hold successful meetings. Most suggestions involve pain-avoidance. In a *New York Times* interview, startup funder Paul Graham described the ideal meeting this way:

> *There are no more than four or five participants, and*
> *they know and trust one another. They go rapidly*
> *through a list of open questions while doing something*
> *else, like eating lunch. There are no presentations.*
> *No one is trying to impress anyone. They are all*
> *eager to leave and get back to work.*

To be fair, not everybody thinks the ideal meeting would be conducted with a mouthful of food. Meetings are the one time where face-to-face interactions can be experienced in real time. Indeed, 80% of people who initiate meetings think meetings are productive and worth salvaging. Their advice isn't to eliminate them but to upgrade them, using the behavioral sciences as a guide for improvement.

We're still going to use behavioral sciences as a guide, and we'll still talk about making meetings more productive, but first we must deal with the literal virus in the room, COVID-19. A tiny microbe has done the one thing almost 200 years of American capitalism could not: alter how meetings are done.

The change may be more indelible than first thought. Through the course of the pandemic, many businesses switched to meeting remotely. Exactly what that will mean for our work lives is still an open question. But it's clear these extraordinary social disruptions will not just be a peculiar characteristic of the *annus horribilis* of 2020.

Working from home

After I finished watching the YouTube clip of the interview, I remember thinking, "I've seen the future—and the future is *funny!*"

You've probably seen the clip too: Professor Robert Kelly, an expert on the Koreas, was being interviewed at home by the BBC. Kelly's children decided to become part of the broadcast and, eventually, internet lore. First up was a toddler in a yellow shirt, who opened the door to Kelly's study and marched up to the computer's camera, smiling and dancing. The interviewer alerted Kelly to the interruption when Kelly's nine-month-old son, harnessed in a circular baby-walker, suddenly rolled in right behind. The scene closed with Mom attempting a rescue, fumbling to get the kids out of the shot, books tumbling to the floor. Even most Hollywood slapsticks aren't this good—or prescient.

There are elements to this video that seemed to foretell the future of meetings. Consider the financial savings of this *in situ* interview. Kelly lives in Seoul, South Korea. Flying him to London for the interview would have been much more costly than simply connecting at home for a video chat. Not having an office commute reduces transportation costs for many businesses.

Another relevant savings factor is morale-based. Workers—Kelly being the possible exception—tend to like working from home at least part of the time. One survey found that only 14% of people sheltering in place during COVID-19 wanted to return to an office-every-day model when epidemiologically possible. Almost half said the best situation would be a mixture, working from home for much of the work week and coming into the office occasionally.

A final savings factor involves productivity issues. Some companies, including industry giants such as Cisco and Microsoft, now report large increases in productivity from their stay-at-home employees. Executives, managers, and employees alike have expressed amazement at the number of meetings that weren't actually necessary to get things done, though the acclaim wasn't universal. Working from home seemed best suited to those participating in a knowledge-based economy, which, as you probably know, is most of us.

Bottom line? Working remotely is here to stay, which means remote meetings are here to stay.

What do we know about Zoom-type meetings? What do we know about designing the home office rooms in which those meetings must take place? Are they all blessings? curses? blends of both? Studies of these issues are just beginning, and the preliminary evidence is in: officing from home is a mixed blessing—and sometimes a *really* funny one.

Vision takes brain power

Let's start with the negative.

I don't mean to pick on Zoom. Other videoconference platforms—such as FaceTime, Skype, Microsoft Teams, and Google Meet—are out there too. What they all have in common is that the brain hates them. Or perhaps, more accurately, it hasn't had a lot of time to adapt to them and still deludes itself into believing it's in the Serengeti. This delusion underlies most of the issues brains have with video chats.

One of those issues is energy consumption. Videoconferencing is an energy-suck. The experience is so prevalent, it's been given its own name—and here we pick on one platform again: Zoom fatigue.

Why fatigue? Part of the fatigue concerns the visual nature of the videoconference. Almost half the brain is devoted to processing visual information. Video information taxes brain resources in ways that other information—such as audio information—does not.

The other part concerns nonverbal information, which the brain's visual system is also built to detect. Zoom and other video-conferencing platforms have either too little or too much nonverbal information, depending upon whose research you're examining. Because remote-tech is mostly composed of faces, it conceals important social information from the rest of the body, leading to some distortion: to cope, you begin inferring things that may not actually be there. You start overvaluing someone's verbal cues, for example, since that's the only other source of sensory information available to you. This compensatory behavior is also exhausting.

Stanford researcher Jeremy Bailenson believes just the opposite can also occur. Depending upon the size of the meeting, Zoom-tech may give people too much nonverbal information. He calls it "nonverbal overload." This overload happens because video chats often include multiple participants—cue the Brady Bunch opening credits here—each person broadcasting their own nonverbal, stare-at-you cues. That's too much nonverbal material, an overload.

Whether too little or too much information, remote meetings create an energy-zapping environment. They recruit half your brain to exercise two of the most fuel-consuming activities: processing visual data and figuring out how to interact socially. It's fatiguing. It's *Zoom* fatiguing, especially when all you have for communicating is your face and a few words.

You can see evidence for this drain simply by looking at what happens in a typical chat. Many meetings collapse into conversations between primarily two people, everybody else spectating.

Is this collapse Zoom-specific? In typical, nonremote business meetings, a four-person gathering is *usually* dominated by two people. Increasing the meeting size to six adds a third person to the airwave encroachment. But nobody suffers from Zoom fatigue in these meetings because, well, there's no Zoom. With remote meetings, you add the annoying energy-drain of continually reediting and reinterpreting the conversation. Given this unpleasantness, it's possible, even likely, that the dropout rate is greater. You begin questioning the point of having multiple participants, especially if it all distills down to an interaction that might have been more easily executed by simply picking up a phone, no videocam in sight.

Unnatural gazes

There's another reason the brain is uncomfortable during video chats. This one concerns the extreme unnaturalness of such interactions.

Consider that remote communication involves faces gazing at each other for sustained periods, which would have been a big deal out in the Serengeti. Sustained gazing is designed to command attention in social mammals. It allows the brain to swallow buffet-sized volumes of social information in a relatively short amount of time. It also takes ridiculous amounts of energy to maintain. In the flesh-and-blood world, conversations are never staring matches, but in Zoom, that's practically all you have.

The timing of a *natural* gaze has been measured. If someone averts their gaze from you within 1.2 seconds of the initial encounter, you begin thinking they're ignoring you. If they stare at you for more than 3.2 seconds, you begin feeling uncomfortable, wondering if something creepy is going on. Getting the balance right is so endemic to who we are as a species that alterations in gazing behavior are considered a mental health issue. In babies and toddlers, avoiding eye contact is an initial sign of autism.

In Zoom-world, this all gets upended. People stare at you during a video chat, and you stare back at them for minutes at a time. Much of the time, you can't even tell if other people are actually looking at you, which leaves you at a total loss when it comes to reading onlookers' reactions. You can end up averting your gaze from someone, not because you're dismissive but because you're looking at the videocam incorrectly.

Another aspect of this unnaturalness concerns the relative size of the human face during these meetings. A typical video chat involves the head filling the screen. But assessing the size of someone's face has deep evolutionary significance.

Why? In our savannah sojourn, the only time our heads perceived a big face was when we were physically very close to another person, so seeing a big head immediately lights a cerebral proximity sensor. Only a couple reasons for such closeness exist with hunter-gatherers: you are either about to engage in close physical combat or you are about to have sex. The brain knows these events aren't going to happen in a remote meeting, but the Serengeti-forged subconscious alerts are still triggered, so the organ must continuously insert editing comments to keep its evolutionary concerns at bay. The brain is so uncomfortable with big faces that its owner's body begins flinching in real time. Yes, *flinching*. Video chats are about as natural as nerve gas.

Additional face-related weirdness comes straight out of Greek mythology. You might recall the story of Narcissus, the offspring of a river god. Narcissus was said to be so gorgeous that when he viewed his reflection in the water, he fell in love with it. So much so, he never stopped looking at his own reflection. The myth says Narcissus eventually died there, so infatuated was he with his own image. Unsurprisingly, we get the word *narcissism* from this myth.

You may or may not have the face of the Greek god-child, but science shows that you certainly have his preoccupation. If you can see your face anywhere in your visual field, you'll pay an inordinate amount of attention to it, selecting it even amongst a sea of other people's faces. Research also shows that it's harder for you to disengage from your gaze once you've noticed your face looking back at you.

Such self-encounters never occurred in Serengeti-world, of course, except perhaps briefly at some waterhole. But they do happen in video chat world, and no one in the room is going to call this natural. Which is the point. Zoom, without face-blockage, can be a very distracting way to communicate.

Work-arounds

Taken together, Zoom doesn't paint a very pretty picture. It's exhausting, impoverished, and unnatural—but it's here, and probably to stay. Which means we need work-arounds designed to minimize the negative effects of the video chat world.

My first suggestion bluntly addresses Zoom fatigue: don't video-conference every meeting. Phone calls, much less exhausting, should be sprinkled into your daily communications. You might consider following an interleaved model: do one video chat, followed by a break (bathroom, food, a bit of exercise—whatever constitutes an interruption), followed by a phone meeting, and repeat this rhythm throughout the day. If that rhythm isn't always possible, you can minimize the cognitive battery drain by simply turning off the video option. You could even establish in meeting invitations that a given meeting will be audio-only, so everybody switches off their camera and experiences what's essentially a large conference call.

I also suggest modifying the on-camera airtime in meetings where videocams aren't optional. Bailenson describes meetings where only the person talking has his or her camera active. Everyone else is simply audio-only, akin to video streaming. He relates that when he uses this model, he gets relief from his usual Zoom fatigue.

Finally, I suggest practicing a few techniques that improve the social dynamics within the videoconference itself. We discussed, given the paucity (and distortions) of the information stream, that misunderstandings and misinterpretations are more likely to occur in video chatting. One way to beat back some of the confusion is by using formal perception checking. You verbally repeat the information you think you heard, then ask for clarification—even if it feels awkward doing so. This kind of formal perception checking improves clarity and understanding even in face-to-face meetings. In video meetings, it's that much more important to check in and make sure you understand exactly what is being said.

Another habit you should incorporate in video calls is an effort to equalize participation. If you've not heard from someone during a meeting, it may be a good idea to gently direct the conversation to the silent party by checking in: "We haven't heard from you for a while. What do you think about what you just heard?" and then wait for a reply. These seemingly artificial habits can quickly become normal if you deploy them consistently.

All these suggestions concern conduct within and around meetings. They are attempts to compensate for some of video chat's inherent weaknesses. But what about the structure of the meeting itself? Is there an organizational design that improves productivity, especially if the meeting is weighed down with the unnaturalness of video chatting?

Possibly. A meeting design that shows potential in improving productivity, efficiency, and clarity, both over video and in the "meatspace" has emerged. Oddly enough, its design originates with a mistake made by some of the brightest people on the planet.

The problem with MOOCs

The "brightest people" I'm talking about are the faculty of MIT (and, eventually, much of the world of higher education). In 2010, perhaps besotted with the potential of digital learning, the MIT faculty decided to put all their coursework online. They called these digital creatures MOOCs, short for Massive Open Online Courses.

The reasoning behind MOOCs was simple: for too long, only a select few students had possessed continuous access to the greatest minds in the world, and only when they matriculated at the country's elite universities. With MOOCs, all of that could change. Even tests could be taken online. The only price of admission would be a sturdy internet connection.

Like many other promises of disruption due to the internet at the time, MOOCs were the advent of something exciting. In the following years, many universities followed MIT's lead and posted MOOCs of their own.

Naturally, people were interested in understanding if MOOCs worked. After almost a decade of study, the jury reached a verdict. In 2019, *Science* magazine published a paper ominously titled: "The MOOC Pivot. What happened to disruptive transformation of education?" It wasn't flattering. Researchers noticed students who took one MOOC course seldom took another. (Only 7% of the kids who MOOC-matriculated in the fall lined up for seconds next term.) When researchers examined students' progress while taking the class, the view got even worse. Only 44% actually completed the *first assignment*. Less than 13% got through one full course.

But there were some silver linings. Very little in the research world is this monolithic, and fortunately for MOOC, these damning data eventually gave way to more nuanced findings. Researchers discovered some MOOCs transferred information fairly well, as long as they conformed to certain rules. These rules were surprising, because they did not follow a traditional lecturing template. Here are the two biggest factors for making MOOC magic:

1. *Prepare in advance.*

The most successful MOOC faculty gave their lecture notes to students in advance of the online experience. Doing so allowed the kids to interact with the material prior to class, question the ideas, discover and characterize confusing parts.

2. *Discuss instead of lecture.*

After this prep step, the MOOC could be taken in real time, but now, instead of the professor livestreaming a lecture, the professor livestreamed a discussion—a structured question-and-answer experience. Common points of confusion could be easily addressed, and a space for lively interactions created. The best parts of college could be experienced not by having "sage on the stage" but a "guide by the side."

MOOCs and business meetings

Can the rules that made MOOC magic also apply to the business universe, especially Planet Zoom? I believe they can, though I am sadly saddled with the word *believe*. Until rigorous, randomly controlled studies show the relevance of MOOCs to Zooms, I'm left using phrases like "these data suggest ..."

What they suggest is something radical, involving a preparation activity followed by a three-step protocol. And this approach may be not just for Zoom meetings but also for any meetings of the future.

The preparatory step is cognitive. As the "showrunner" of a meeting, your first job is to make crystal clear *in your own mind* what the meeting is supposed to be about—a cognitive mini-mission statement. Write that mission down in a sentence. Then construct agenda items with one ground rule: nothing varies from the mission statement. It's the equivalent of outlining a lecture.

If the idea is to leave meeting participants with memorable takeaways, it's wise to follow a few rules for agenda construction. Human beings retain information better if the goods are presented in a hierarchical, gist-to-detail fashion (with typical retention rates being 40% better). If an agenda item involves revealing a spreadsheet, the participants should be told why the spreadsheet is important (the gist) before they see the spreadsheet (the detail)—general information first, followed by specifics.

Once you are finished with your prep work, you now have in your hands a clear, concise document that will serve as the backbone of the meeting. We'll call this document a "showrunner's agenda."

Now you are ready to move to the three-step protocol:

1. *Deliver your showrunner's agenda in advance.*

The ideal time to send out your agenda would be a day or two before the actual meeting, including any supporting visual aids (such as slide decks).

2. *Have attendees pre-read the information.*

The participants examine the received materials and make lists of their questions, comments, and requests for clarifications. They mark up your documents with abandon. Then they bring their lists to the video chat, prepared to discuss whatever is on their minds.

3. *Begin the videoconference.*

You fire up the videoconference, starting the get-together with one critical difference: You're not going to lead a meeting. You're going to lead a discussion. Rather than storming down a punch list, you'll listen to questions and concerns as thoughtfully as Zoom allows. When I do remote lectures, I briefly summarize important agenda items then, as quickly as possible, turn the lecture over to the audience by saying:

"Okay, folks, the next part of this show is brought to you by you. Are there any questions? How may I be helpful?"

It only takes a few awkward minutes before a boring MOOC turns into a full-fledged, I'm-still-paying-attention learning event. That's despite the thinness of the communication atmosphere surrounding Planet Zoom.

Why home offices are in our future

Our previous sections concerned videoconference-based meetings—why they are so exhausting and how to navigate them successfully. But we haven't yet talked about the current setting for these newly popular meetings: the home office.

Home office meetings have wiggled themselves into a probably permanent position at the business table. We're shortly going to discuss why that's true, then pivot to discussing what those offices should look like. Behavioral neurosciences can play a surprisingly robust role in home office design. For now, let's examine the word *permanent*.

As the COVID-19 outbreak continued, home offices became a stopgap alternative. Some CEOs loved this alternative, encouraging all their employees to consider permanently working remotely. Other CEOs hated having a remote workforce and couldn't wait to come back to the previous model. Merrill, mentioned at the start of this chapter, was one of them. It will probably take years before the workplace settles into a new equilibrium, with CEOs squabbling all the way.

Sadly, even their points of view will have to be held loosely, regardless of our success in beating back COVID-19, since viruses have many deadly, pandemic-friendly cousins waiting in the wings. COVID-19, in particular, has shown to be very good at mutating. Variants, some more contagious than the original strain, have surfaced all over the world, including here in the United States. With globalization a perpetual fixture in the world economy, new viruses will circulate faster than ever before. Indeed, since the 1980s, the number of worrisome infectious microbial emergencies has tripled.

One reason some researchers feel the home office alternative may be long-lasting is its money-saving potential. Working at a home during the pandemic at least some of the time meant you weren't working at an office at least some of the time. The implications were logical: You didn't always need a big office building to conduct big-office business. Heck, some companies didn't need a building at all. One study published in the *Harvard Business Review* put the savings on furniture and office space at $1,900 per employee. Coupled with changes in transportation costs, the ability to offshore work to people's dens meant the potential for recovering big bucks—catnip to anybody familiar with even rudimentary spreadsheets.

Another reason some researchers believe strongly in the indelibility of the home office concerns employee productivity, which surprisingly showed mixed results during the pandemic. In places where it's been studied carefully (the pandemic made a terrific pre-post research program), most companies reported no change in productivity.

Coupling any of these findings made most companies salivate. There are compelling reasons for working from your den or your closet or spare bedroom. Which means home offices, like debt-laden college students, won't go away any time soon. Exactly how long they stick around won't be known for years.

Executive function defined

Given that the future of your work involves some of the same places where you write your checks for rent or mortgage, what should a home office look like? And *act* like? Can brain science help?

It might. When people fail at designing their home office, it's usually because they don't pay enough attention to a cognitive gadget called *executive function* (EF). We need to spend a few minutes ¹efining EF before getting into the weeds with the look-and-feel ˙ of home offices.

Let's start with an illustration from the movie *Saving Private Ryan*. Central to this movie, one of the most realistic depictions of combat ever filmed (some vets ran out on the movie after the first thirty minutes), is Tom Hanks's character, a captain in the US Army. We see battle through his eyes. He experiences plenty of horror, at one point dragging a half-blown-up corpse up Omaha Beach. But he checks his impulses, assesses the situation, finds men he can work with, and starts barking orders. With the battle raging all around him, he organizes an assault on a German gun emplacement, ultimately seizing his target and achieving his objective.

Hanks isn't playing a superhuman. (His first scene shows his hands shaking as he prepares to land.) What the character had—illustrated in the most powerful and horrific way possible—was robust EF.

EF is often crudely defined as the behavior that "allows you to get something done." In more scientific terms, it embraces two suites of behaviors. The first is *emotional regulation*. This type of regulation includes behaviors like impulse control, which researchers call "inhibition." Hanks's ability to carry on amidst the gore, though every cell in his body wanted to hide, is a prime example of emotional regulation. The second suite, called *cognitive control*, includes goal setting, the ability to plan something independently and provide a framework for it with little assistance. Cognitive control also includes the ability to focus, get distracted, then refocus again at will. (People with ADHD often lack this cognitive gizmo.) Cognitive control allows your brain to organize disorganized inputs into something more manageable—instilling order—often by using a hierarchical gist-to-detail system.

Working at home successfully requires a number of gadgets in the EF toolkit. And oddly enough, the physical design of the home office can help maximize the functioning of those gadgets (including running productive meetings). We'll start with advice about how the office should look, then move to how people should function in it.

Trading spaces

The first suggestion may be the hardest. The best way to harness productivity while homebound is to dedicate a space where you will office. It's a single-use, I-don't-do-anything-else-in-that-space place. A room with a closeable door is ideal; indeed, the discussions that follow assume one. I recognize that such spaces may be impossible in some homes. For those of you who can't dedicate an entire room of living space to an office, designate a temporary single-use area for the day—a closet, a space on the dining table, a quiet corner—as a work-around.

Why dedicated space? Breezy articles about home-officing often emphasize the importance of creating psychological boundaries. Because we are performing more and more disparate activities in one place, the boundaries between those activities are in danger of dissolving. And that, as research shows, can produce psychological problems. Annihilation of the armistice in the work-life balance battle is a major casualty, much of it wrapped up in a notion called *self-complexity theory*. This theory examines the effects of context on an individual's multiple social roles.

Researchers have found that, to be healthy, humans need a wide variety of social settings and contexts. These settings and contexts need to be kept separate—boundary-dissolving isn't healthy. Researcher Gianpiero Petriglieri puts it like this:

> Imagine if you go into a bar, and in the same bar you talk
> with your professors, meet your parents, or date someone,
> isn't it weird?

This lack of boundaries isn't just weird. It's also potentially destabilizing. If you can't separate work life from home life, you are at greater risk for burnout. You become more vulnerable to certain affective disorders (such as depression and anxiety) simply because boundaries between your social roles become increasingly hard to manage.

Most of our social roles are supposed to happen in different places, according to Petriglieri. Siloing off a portion of our living space for work helps us slip into "work mode," a mindset marshalling the suite of behaviors and activities devoted to your job (as opposed to "home mode," in which you shed the armor of the workplace and focus only on matters of the home.)

Mode is a squishy term, in my view, but the core idea is based on well-established behavioral science. Years ago, researcher Alan Baddeley discovered something termed *context-dependent learning.* He had people learn something—often lists of words—in a given physical space. He then asked them to recall the words hours/days/weeks later. He found that retrieval was always more robust if the learners retrieved the information in the space where they originally learned it. He experimented with many contexts, including memorizing lists while in a wetsuit under water!

Turns out, the brain is really good at recording the physical contexts in which intellectual activities take place. It uses that recording to assist in recalling the activities expected of it in that space. The effect is so powerful, it's been used in sleep research. If you have trouble sleeping, researchers suggest you dedicate a room with slumber as its single, ironclad use. When your brain encounters the space, it says to itself, "Oh, this is where I usually sleep. Therefore, I will get drowsy." This technique is very effective for people with long-term sleep issues—and perhaps for people needing productive home offices.

Context-dependent designs have collateral benefits. When you create dedicated spaces, you cut down on distractions, for example. A single-use space also allows you to leave something at quitting time, and then, next day, return exactly where you found it.

How does creating a home office improve meetings? Recall that the world of videoconferencing already exerts artificial pressures on the brain. Closing the door so you can concentrate on the less-than-optimal incoming information becomes critical. Since focusing is a card-carrying member of Team Executive Function, you're summoning the one cognitive gadget capable of making you more productive on Planet Zoom.

Schedule control

While the first piece of advice concerning working from home involves creating dedicated workspaces, the second involves answering this question: What should you do once you sit down at your desk? That's a tough query, but it has a simple answer, describable in two sentences:

Make a schedule. Stick to it.

The research world formally calls this *schedule control*. It's contrasted with another experience researchers study, *job control*, or the study of how you work. Whereas *how* you work can be hard to characterize (we'll discuss it later), schedule control is not. You create a punch list, usually filled with tasks: *I'll do X task during Y hour for period of time Z*. Boring as a spreadsheet—and as vital.

The linear idea of schedule control is, I realize, hard to reconcile with the uncomfortable fact that life's often messy. Moreover, tasks don't always lend themselves to being accomplished in tidy, predictable aliquots of time. That's especially true working from home, where interruptions from colleagues swap places with interruptions from family. But it's important to exert control, however hard that becomes.

One way is to divide your efforts into small, easily consumable goals, a strategy that writer Anne Lamott employs when penning her books. She picked up that habit during her childhood years. In her book *Bird by Bird*, Lamott recalls how her younger brother once tried to write a book report about birds. Overwhelmed by all the avian species he had to research, he started crying. Lamott's father came over and told him, "Bird by bird, buddy. Just take it bird by bird."

ïding tasks into what she calls "short assignments" transforms projects into bite-sized chunks, rendering even the largest tasks into readily digestible pieces.

Research shows that if you don't practice schedule control, you become less productive. Not surprisingly, there's a brain-science reason why. It involves the rocky relationship between negative stress and executive function. We'll give a more full-throated treatment to negative stress later in this book, but here's the bottom line: For most people, it isn't the presence of aversive situations that causes negative stress; it's the inability to *control* the aversive situation. The more out of control you feel, the more likely you are to experience negative stress.

Why is that point important? Research shows that negative stress can rust many of the gadgets in the EF toolbox. And that has consequences. Executive function participates in most of the activities involving productive scheduling: planning, oversight, focusing/defocusing/refocusing, impulse control. Negative stress cripples areas in the brain involved in establishing and maintaining those functions. We even know the hormones doing all the damage.

It follows then that the more control you feel over your schedule, the less likely you are to experience negative stress. And conversely, the less control you exert over your activities, the more likely you are to experience tension. Which puts you on a scheduling treadmill from hell. Negative stress from a lack of schedule control will hobble EF, the very gadget which would normally allow you to churn through your day in the first place.

The solution is so easy but so important that I feel I should repeat it: Make a schedule. Stick to it.

Procrastination

How this relates to our discussion about meetings is obvious. If you stick to a schedule, you can insert your teleconference into a specific time slot and set up the parameters, with start and end times. If you keep to a disciplined commitment, chances are you'll be more productive and avoid working longer hours.

Not that there aren't enemies to this disciplined approach. One of the biggest has nothing to do with timed meetings, or even other people. It has to do with yourself—especially if you're burdened with one of the most emotionally freighted words in business: *procrastination*.

Procrastination can be thought of as a war fought on the battlefield of productivity, a contest between what you need to do and what you want to do. The unusual outcome is that it's almost always the battlefield that loses.

Researchers have studied procrastination. Findings show that procrastination is not due to a lack of discipline, though most procrastinators feel otherwise. What fuels procrastination is an attempt to avoid adverse feelings. If you have a habit of evading things that bug you, procrastination is probably your constant companion. And if you think that sounds like a lack of impulse control, one of the two great pillars of executive function, you are right on the money.

Given that procrastination is in EF's wheelhouse, people must change the way they handle negative feelings in order to beat it. For the brain, handling negative emotions always requires excess energy. Yet research shows that procrastination is more likely to occur at low-energy times of the day, which for many people is midafternoon. If that's true, don't put off emotionally difficult tasks for a time when you've got the least amount of energy. Determine if a given task is tough, then schedule time to deal with it when you're at your peak. For many of you, that will be in the morning. Or after your eighth cup of coffee.

You should also insert more energy into your day. Perform activities that increase the impulse-control component of executive function. While that sounds easier said than done, we actually know a lot about how to do this. Research says that to spend more time being productive, you need to spend less time working. The evidence says that sometime during the day, you should take a nap, then go for a run.

Of bicycles and beds

Actually, the research doesn't really care about the order. It simply demonstrates a need to insert both activities into your daily work. Both behavioral hygiene habits improve brain function, specifically executive brain function, which helps you to better navigate the hazards of remote meetings.

The exercise needs to be aerobic in nature, about thirty minutes every twenty-four hours, preferably a half-hour run that gets you out of the house. Why? Regular aerobic exercise improves nearly every measurable aspect of executive function in nearly every age group ever studied. One inquiry examined a truly tough group, populations with mild cognitive impairment. This research showed that exercise improved short-term memory, a key component of executive function, by a whopping 47% after a year's effort. We even know *when* to schedule this EF-boosting activity: before 3:00 p.m. That's to ensure a good night's sleep.

Which is the other great habit to insert into your workday. Research specifically suggests naps in the midafternoon, when you're most likely to get drowsy, executed as regularly as lunch. Mark Rosekind, formerly at NASA, showed general cognition improves by 34% in people who regularly nap. The benefits are wide-ranging, from cardiovascular improvement to changes in executive function, particularly the cognitive flexibility/focusing component of EF.

On average, a nap should be no longer than thirty minutes and taken at the ideal time for most people, sometime between 2:00 p.m. and 3:00 p.m. Notice I said *most people*. There's enough variation in the *when* of this science that sleep researcher Sara Mednick developed a calculator for it, which you can find in her book *Take a Nap, Change Your Life*.

Taken together, these data suggest that if you want to overcome the natural drawbacks of remote meetings, you need to up your executive function game. And the way to do that is to create an ironclad, goal-setting structure that includes short runs, short naps, and short assignments.

These suggestions may seem weirdly counterintuitive, but the workplace was never designed to account for the electrical needs of the human brain. It certainly wasn't designed to improve executive function. But the data are remarkably clear. Being successful when it comes to working from home means—to quote a hopelessly hoary aphorism—working smarter rather than working harder. Odd to say, but being more productive at work actually means doing less of it.

THE HOME OFFICE

Brain Rule: *Your workday might look and feel
a little different than before. Plan accordingly.*

- COVID-19 has altered the way we conduct meetings, probably for good. Videoconferencing will play an important role in how you communicate with coworkers.
- Videoconferencing is particularly draining, siphoning more of the brain's energy than if you talked in person. To curb the drainage, occasionally switch your camera off, check in regularly with others on the call, and use formal perception checking.
- To get the best out of any meeting, create a showrunner's agenda ahead of time, send it to participants early as a pre-read with a request they come prepared to discuss the contents, and then conduct a discussion rather than a lecture.
- If you are working from home, dedicate a space (no matter how small) that is a single-use workspace.
- To optimize your workday, make a schedule. Stick to it.
- Procrastination is an avoidance of negative emotions. If meetings drain you emotionally, schedule them when you have more energy—probably in the morning.

the business office

Brain Rule:
The brain developed in the great outdoors.
The organ still thinks it lives there.

I NEED TO CONFESS SOMETHING before we get started on this rule. I harbor a powerful bias toward the work of Harvard biologist E. O. Wilson. The subject of this chapter, which concerns the influence of the natural world on human behavior, is marinated in some of that worship.

Wilson is an unlikely scientific hero. He's slow to talk, blessed with a mild southern Alabama accent—as close to the demeanor of Mr. Rogers as scientists get. Much of his work is observational, which is extraordinary, because he's been blind in one eye since the age of nine. That impoverished vision forced him to study small things up close, which he has done to ferocious acclaim. He's considered by many to be the world's authority on ants.

My hero worship does not spring from his powerful, noncontroversial contributions to insect biology, however. Nor does it come from the science I practice, which is about as far removed from Wilson as Boston is from Seattle. My worship comes from his equally powerful and, in times past, extraordinarily controversial take on human behavior.

Wilson is curious about how nature influences human behavior in our everyday lives. He popularized the term *biophilia*, meaning biology-loving, a word originally coined by philosopher Erich Fromm. Whereas Fromm meant it more as a psychological orientation to explain present human behaviors, Wilson has taken the term on a road trip through human evolution. Here's a quote from Wilson himself, describing the engine that powers biophilia:

> *Human beings are biologically predisposed to require*
> *contact with natural forms. People are not capable of living*
> *a complete and healthy life detached from nature.*

Since this chapter is about workplaces, our focus will be the world of office buildings and conference rooms—and the considerable amount of review these have received as a result of the pandemic. Should there be a return to the typical office? Should most people be allowed to work at home? Should there be a mixture? The notion of working at an office—once the unbreakable, steel-and-girder backbone of American business practice—underwent a considerable tempering in 2020 and 2021.

Given this softening, it's a great time to reimagine the whole idea of the work office, although, as we'll see, most of the pre-pandemic data are still relevant. The first half of the next section will cover what *predisposed* means when humans react to their physical workplaces. The second half will tackle Wilson's notion of "healthy life." We'll discuss what happens if we ignore our Darwinian history when designing office spaces, as well as what happens when we try to bring some of the savannah indoors. As you will see, the office of the future might benefit from its designer looking back at what life would've been like if we set up shop in East Africa, then worked there for nearly 60,000 centuries.

Environmental sensitivity

My last sentence deserves an explanation. Evolutionary biologists explain that our journey toward humanhood began 6–9 million years ago. That's when we diverged from chimpanzees, then followed the road leading to gleaming cities and income taxes. Since modern civilization is a recent invention, we're left with an interesting, unsettling number: for 99.987% of the time our particular species has spent on the planet, we've lived our lives in settings composed of natural elements. We evolved our big, fat, talented brains under conditions that favored surviving grasslands, not traffic jams. Biophilia argues that we've not been civilized long enough to escape evolution's influence. Thus, we still have preferences for natural things.

Parts of these ideas are testable—we'll get to some data later—but there's an important factor to keep in mind as I describe them, one that concerns a certain type of neurological sensitivity. There were times in our six-million-year road trip when the climate grew very unstable. We had to adapt to that instability as surely as we had to learn to fear snakes. We had to become exquisitely sensitive to change.

That sensitivity, amazingly enough, is something you can measure, beginning with the tiny neural circuits in your brain. As mentioned in the introduction, every time you learn something, you rewire those circuits. Quite literally, new connections form, electrical relationships change, and neural circuits strengthen and weaken. It's happening even as you read this sentence.

That sensitivity's built in. Indeed, it's the engine room that powers our ability to adapt, and comes equipped with a curious behavioral consequence: ignorance. We come into this world knowing very little about it, which means we have to learn just about everything. If we weren't sensitive to its lessons, we'd learn nothing—and then we'd be dead.

Not all creatures are born burdened with such a steep learning curve. Wildebeest babies, for example, are ready to run across the Serengeti a few hours after they are born. Humans take almost a year before they can navigate—most with faltering steps—across even smooth, unchallenging surfaces. Yet we survived, adapting and solving problems related to grasslands and the towering heights of the Rift Valley in East Africa. Those who weren't sensitive enough to adapt to our Tanzanian playpen perished quickly.

You can observe our finely tuned sensitivity using experiments that don't take eons to complete. Consider something behaviorists call "priming." One classic experiment involves subjects reading synonyms for *aggression*. They then read (or watch videos) about characters exhibiting neutral or ambiguous behaviors. When asked to evaluate these neutral behaviors, the subjects don't choose words like *neutral*. They invariably chose words related to aggression. Their brains were sensitive enough to have their outer environment prime them for future responses. Use synonyms for *kindness* instead, and you'll find similar changes, only this time in the opposite direction. Everyone is now nice.

We own brains so ridiculously attuned to their outer environment and prone to mold to it, researchers can observe these adaptations in the short term. If we're so sensitive to our surroundings that we'll change perspectives with just words, what must 60,000 centuries have done to us? If you're Wilson, you suspect those years have done a lot. It's time to uncover exactly what "a lot" means.

Change and stress

While Wilson believes we have brains that prefer natural forms, we also have brains capable of creating cities, unnatural in every way. Is our ability to adapt to change stronger than the preferences baked into us by the Serengeti? The answer is an unsatisfying "sort of." We really can adapt to change, but we take enough of the Serengeti with us to feel its interfering influences on us, sometimes painfully so.

Stress is a perfect example. Under proper circumstances, stress responses are real friends to your inner hunter-gatherer. When a threat emerges—say, a lion—your heart pounds, your breathing quickens, and your senses sharpen. Much effort is devoted to pushing blood into your thighs, allowing you to flee at full speed. It's sometimes called the "fight-or-flight response."

If you peer into someone's brain while they're experiencing stress, you'll notice that a network of neurons called the *salience network* (SN) becomes hyperactive. It's a coalition of neural networks that supervise your ability to get out of Dodge when the bad guys come to town. You can think of an activated SN as a flashing red light. This network is responsible for sending out the signals that trigger your physical responses to a stressor: the pounding heart, the quick breaths, the sharpening senses. Sharp observers will also notice an extreme individuality in the intensity of this activation.

Interestingly, there are natural governors to our stress responses, designed to ask, "Can I turn myself off now?" These queries happen almost as soon as threat responses commence. The biggest brakeman? The stress hormones originally recruited to supervise the muster! One such hormone, cortisol, is part of what's called a *negative feedback loop*. Once the red alert is initiated, cortisol immediately asks the brain when it can turn things off, beginning with its very own manufacturing site.

The reason's simple: the reaction is so energy-draining that the system is wildly in danger of collapsing if it's pushed too far. That's why researchers believe most threat responses were aimed at solving short-term problems. The lion either ate you or you ran away from it, but the threat lingered for minutes, not years.

And it's here where we run into a big, fat problem. In modern life, threats *can* last years. In the twenty-first century, you can be stuck in a job you hate for decades. You can be stuck in a relationship you hate for decades.

Even if you love what you do, some jobs are so stressful that your cardiovascular system becomes injured. Your immune system can too, meaning you get sick all the time. You can easily push the system too far, and not just once but over and over again. The red light is only meant to stay on during emergencies; it isn't meant to keep flashing for hours, days, or years on end.

Red light, green light

This pattern of repeated stress is so prevalent in modern times that it even has a name. *Role overload* is the term researchers use when your day is routinely filled with tasks you can barely complete—in other words, when you have too much to do. *Burnout*, also a formal term, is the consequence of sustained role-overload. It's your brain's way of waving a white flag, then finding some corner in which to have a good cry. In true burnout, heartbreakingly, that good cry lasts years.

Both role overload and burnout are part of an overarching experience called *mental fatigue* (sometimes called *cognitive fatigue*). When sustained stress exhausts you, your ability to do your job competently is in serious jeopardy. Error rates increase. Absenteeism becomes common. You get moody and grouchy, your personality inflicting blunt force trauma on everyone you meet. Your risk for depression and anxiety skyrockets.

Mental fatigue is so powerful it can be imaged in the human brain with noninvasive imaging technology. In such imaging, you can observe what happens to the organ just before the collapse occurs: a big ugly red spot of hyperactivity suddenly appears behind the forehead (anterior prefrontal cortex) just prior to the brain shutting down. It looks like the nastiest Doppler tornado warning from a Kansas weather report, if Kansas were in your skull. Keep that up for too long, and the sustained stress causes actual brain damage. The culprits that kill brain cells are literally the stress hormones designed to keep you safe. They've just been pushed to overdrive.

This is obviously a large dollop of bad news. Is there any good news? Are there antidotes to role overload, to burnout? If scientists can image what brains look like under the cruel rule of mental fatigue, can they also image what brains look like when liberated?

The answer to those three questions is yes, yes, and most emphatically yes. When the brain becomes relaxed, researchers observe several things, one of which is surprising. First, when people calm down, many of the brain regions mediating the stress deactivate, which is hardly new news. But when the brain starts to rest—here's the surprise—certain interconnected regions suddenly become quite active. These regions make up the neural confederation called the *default mode network* (DMN). The regions directly behind your forehead (medial frontal cortex) and toward the middle of your brain (posterior cingulate cortex) comprise much of the DMN's real estate.

The irony here is that when you are calm, relaxed, at your most completely passive, and have positively nothing to do, your DMN begins firing on all cylinders. (You have to actively suppress it during times of intense focus.) This act of derepression pushes you into a state termed *indirect attention,* a low-arousal state also called *soft fascination,* which is characterized by the type of languid feelings you get when looking at clouds slowly drifting across the sky, or gazing at fish in an aquarium. You might know it by the more familiar term *mind wandering.*

Whatever the name, this state is characterized by the brain not tarrying on any single subject for sustained periods of time. Instead, it becomes subject to the defocusing, hypnotic, electrical rhythms of your DMN.

As the DMN became more fully characterized, researchers noticed that something called *task-negative* responses led to the activation of specific types of creativity and idea generation, now referred to as *task positive* behaviors. As we'll see in a later chapter, the best way to incite creativity is to stare at goldfish.

That's a big deal, especially in businesses that depend upon creative output. The cure for mental fatigue is not simply to turn off the SN's flashing red light in reaction to stress, role overload, and ultimately burnout. It's to turn on the cool, refreshing green one–the DMN–then watch life swim by.

Attention restoration theory

The default mode network and the salience network are mortal enemies—and not equally matched. Whenever there's a fight (or flight), the SN always wins. If you think of the DMN as that refreshing green light, the SN begins active duty by taking its powerful neurological finger and turning the DMN off.

So, the big questions are: How do you get an overscheduled, stressed-out brain to switch its DMN back on? What evidence-based processes allow people to reengage in their jobs? Their lives? Their worlds?

The answer, as you may have guessed, is as close as your nearest ant biologist. The late Stephen Kaplan, a research psychologist who took Wilson's advice and transformed biophilia into a testable set of ideas, fathered something called attention restoration theory, or ART. Put simply, ART hypothesizes that spending time in environments more closely related to savannahs than skyscrapers is enough to restore brain balance. Here's how one researcher couches ART's central tenet:

> ... the mental fatigue that is associated with a depleted
> capacity to direct attention may be overcome by spending
> time in environments rich in natural stimuli.

Turns out, there's a great deal of empirical support underlying this idea, cheered on by that not-so-small biophilic congregation over at Harvard. You can just hear Pastor Wilson shouting *amen*.

Hints from hospitals

One of the first systematic tests of ART ideas came from researching windows in hospital rooms, no kidding. Researchers noticed that patients recovering from surgery seemed to heal faster—and were less cranky while mending—if their room window looked out over trees. Formal numbers backed up the anecdote. Surgery patients used less pain medication post-op if they gazed at natural environments as opposed to, say, brick walls. These patients also needed less emotional support from the nursing staff. And, to the delight of hospital administrators everywhere, they went home a day earlier too. These results held up even when controlling for age, sex, and exposure to specific doctors and nurses.

The man initially responsible for this work, Roger Ulrich, was eventually written up in *The Atlantic* magazine. Said the article:

> *... patients who gazed out at a natural scene were four times better off than those who faced a wall.*

Yep, *four* times.

I call findings like these "rudder research." That is, while the study subject is modest (trees in a window?), the implications are capable of steering giant research ships into uncharted waters.

Which is exactly what happened. Researchers began looking at other natural phenomena, such as natural light, and found that these produced results similar to the trees. Rooms filled with natural light improved post-operative recovery times for back surgery patients (22% fewer painkillers, a 21% reduction in medical costs). Patients staying in sunless ICUs recovering from myocardial surgery had an average length of stays 43% higher than patients in sunny rooms. Natural light even affected mortality rates in those ICUs. The rate for male patients in sunlit rooms was half the rate of those in sunless ones (4.7% vs 10.3%).

Natural trees and natural light changed people's relationships with stress, even under the severest, most uncomfortable conditions. And not just in surgical suites.

Psychiatrists have known for years that light affects mood, but even in severely troubled minds, the natural world penetrates deeply. You can read papers with titles like "Sunny Hospital Rooms Expedite Recovery from Severe and Refractory Depressions" as examples. Even the type of light can affect recovery. Bipolar patients assigned to rooms facing east (where patients observed the direct light of sunrises) stayed in the hospital on average 3.7 days fewer than patients in rooms facing west (where patients observed dusk).

The brain really, really, really wants to be outdoors. This has really, really, really big implications for how hospitals should be designed. The natural world has de-stressing talents of such strength that they can even change people's relationships with painkillers—and death.

More hints from the natural world

Robust as these medical data appear, are they transportable to other nonmedical enterprises? After all, most of us thankfully don't spend the majority of our time in hospitals. Do the de-stressing effects of the great outdoors apply to those of us used to staying in the great indoors? The answer is yes. The gravitational pull of the Serengeti is strong enough to reach us as we hop from one indoor setting to another, coaxing us to come out and play.

Let's begin in our living space before the workday even starts. Researchers from the Netherlands found that people who lived near natural environments started their days with less mental stress than people who didn't. They had less depression, fewer migraine headaches, less heart disease, and, oddly, fewer allergies. By "near," the researchers meant living within about 3 kilometers of greenery. UK researchers noticed similar mental health benefits, even after controlling for employment stability, annual salary, and education (and they monitored where 10,000 people had lived over a period of eighteen years). And *National Geographic* described an even larger international study, this one tinged with economic questions. The study showed that the health benefits to a person living near green spaces were the equivalent of getting a $20,000 raise.

But when I say "green spaces," just what do I mean? What elements are green spaces composed of that make them so beneficial to both our physical and mental states? I'd like to expand our discussion of the world of green by actually talking about the *color* green, using a movie which hardly had any green at all, except for one brief glorious moment.

What we talk about when we talk about green

"Can I cook or can't I?"

That's an actual line from a movie, the famously sexist remark uttered by actor Bibi Besch in *Star Trek II: The Wrath of Kahn* to Admiral James T. Kirk. Her otherwise unbiased character, Dr. Carol Marcus, is the inventor of the Genesis Device, a machine that can conjure up Gardens of Eden from barren, lifeless rock. She made her remark just before showing off the device's handiwork: achingly green jungles, lace-like waterfalls, cliffs, gossamer streams, sparkling lakes, a light source. For 1982-era special effects, the visual impact is still quite remarkable. Kirk was enchanted. So was I. Years later, I used it in a graduate seminar lecture. Now it's even found its way into this book.

Why my emphasis? As with much good science fiction, the scene was prophetic. Dr. Marcus's garden is twenty-first century biophilia dressed up in twenty-third century clothes. It has all the green elements known to improve health in humans. We know these green elements improve human health because researchers have been testing aspects of biophilia around the world for over a decade.

The first biophilic experiments measured what happens to your behavior when you walk through some forested area—then compared it to your behavior when you walk through an urban setting. Researchers from the UK found that if you take a walk in the woods for even a short period of time—I'll discuss how much later in this chapter—your behavior begins to change. The researchers used a psychometric test called a Moodiness Index to measure sylvan effects versus urban effects. Scientists using the index to measure tension, anger, confusion, depression, and fatigue found that subjects who took a walk in the trees reduced all these negative feelings. The effect is multiplied if your stroll includes water, as one might find in a stream or waterfall. They've even given it a name: *green exercise.*

There are medical professionals pushing the National Health Service to prescribe doses of it like aspirin.

Similar researchers halfway around the world (Chiba University, Japan) revealed the same thing. Strolls through their forests, when compared with strolls through their cities, will give you a 12% decrease in stress hormone levels (cortisol), a 7% decrease in nervous system activity, and a 6% lowering of your heart rate. Like the Brits, they also showed a decrease in depression. They gave the activity a much better name: *forest bathing*.

Confirmatory results were also obtained by researchers in the United States. One group even focused on the color of the tree canopy under which a person walked (orange, yellow, or green). Every pigment produced calming effects, but the biggest effect occurred when the canopy was green. The de-stressing effects weren't small, as measured by skin-conductance responses (a way to quantify stress in real time). A green canopy reduced stress almost 270% more than a yellow one.

Dr. Marcus would approve, I sense. Her intellectual heft included, after all, a lot more than just meal preparation.

The physiology of green

It appears the natural world—particularly the color green—exerts calming effects on us like a good therapist. But which parts of the body get the counseling? Some of green's gentlest effects involve something called the parasympathetic nervous system. To understand how this works, you need to know something about that system.

We can divide most of the neurological electrical cabling coursing through your body into specific systems, organized something like nesting Russian dolls. The two largest divisions are the central nervous system (which is the spinal column and brain) and the peripheral nervous system (which is everything else). The peripheral is itself divided into two parts: the somatic system and the autonomic system. The autonomic system is then *further* subdivided (see? Russian dolls!) into sympathetic and parasympathetic branches.

Got all that?

When I previously mentioned the fight-or-flight response, I was describing the stimulation of the sympathetic system. I didn't mention its parasympathetic counterpart, however, which supervises a much more delightful suite of experiences: calming things down after stimulation. Parasympathetic behaviors are sometimes called "rest-and-renew" behaviors. The supervisor of both is the salience network, the previously mentioned network of threat-responding neurons.

Scientists suspect that forest bathing triggers responses from both the sympathetic and parasympathetic systems. Exposure to natural elements lowers your stress hormone levels, as first mentioned in the Japanese work. This response indicates that trees are telling your sympathetic nervous system to "shut up." Nature simultaneously sends a kinder, more mellow invitation to your parasympathetic system: "Do your thing." Forest bathing coaxes your body to start storing energy rather than expending it. Blood vessels begin relaxing (dilating), the heart rate slows, and digestion increases (that's why energy supplies get restocked). People who forest bathe report that it makes them feel rested and renewed, as with any parasympathetic stimulation.

This positivity is directly observable by measuring recovery times, the speed at which your body returns to a peaceful equilibrium after experiencing stress. The natural world cheerfully shortens this time, measured anywhere from the sweat on your skin to the rate of heartbeats in your chest. Even the central nervous system, the one composed of your brain and back, gets nudged along for the calming ride. In response to trees, blood shifts to brain regions associated with empathic behaviors, self-awareness, and goodwill (precuneus, insula, and anterior cingulate). It also shifts away from regions that control your strong passions, emotional responses, and memories of those emotional responses (amygdala and hippocampus).

The importance of green

These effects—calming as a bubble bath—obviously have implications for workplace design, particularly in high-stress occupations. Indeed, creating calming green environments may be the clearest what-to-do-next-Monday suggestion the cognitive neurosciences can make toward workplace engineering. And here's the interesting thing: I really do mean *green* environments, as in the color green (wavelengths hovering around 556 nanometers). Remember the canopy data and how effective the actual color of green was at de-stressing people, compared with other pigments?

The data have gotten more granular these days. And one of the most interesting details these data show is that you don't have to take walks around the great outdoors to de-stress. You can also take walks around the great indoors, as long as you can sojourn amongst indoor greenery.

Does that mean office plants? Yep. Exposure to those lovely nanometers can be carried on the backs of office plants, literally, which is where things get as practical as dirt. People in an office filled with plants experience productivity boosts of 15% over controls and become less fatigued in the process. They also, oddly enough, get less sick. We even know the reason why: it's because those plants pass gas.

Plants emit volatile oils and gases, some of which you can smell. (Can you recall the odor of a forest? Those are volatile gases.) Scientists isolated a subclass of these gases called *phytoncides* and have been investigating their practical uses for decades. It's been a fruitful enterprise. Whiffs of phytoncides have been shown to boost a group of cells in the immune system with the sinister name *natural killer cells*, or NK cells. The name might sound awful, but you actually want the numbers of these cells to increase: NK cells target viruses and tumors.

Green's boosting power isn't small. You get a 20% increase in NK cell populations when you're exposed to office plants and a whopping 40% if you take the whiffs outdoors. It stays at those levels for seven days. And even thirty days later, there is still a 15% increase over controls. Though the test plants were cypress trees and their accompanying oils, most other green plants make these immunity-boosting products. Which means most offices should have green plants and lots of them. It's not a stretch to say that most offices should look like indoor arboretums.

Green and blue and natural light too

Green has other talents besides arming immune systems and creating calming canopies. Green has the additional ability to focus the mind like a magnifying glass. And in a dose-dependent fashion—which means the more that wavelength hits the eye per unit time—the better the focusing becomes. The effect is even powerful enough to change behaviors in populations known for not being able to focus on anything—young people with attention deficit disorder.

Given how Darwinism relates to color, it's no wonder. Consider that the African savannah was not a region known for being awash in the color green. Nor was it awash in water like a rainforest, except during certain seasons. As a result, any sudden encounter with lush green color would have been not only unusual but also probably exciting, signaling the possible presence of life-giving water close by. Researchers speculate that if our ancestors could focus on this photosynthetic fact, they had a better chance of taking in another day, or at least another drink. We humans may have learned to concentrate on green simply because we so often had to concentrate on thirst.

The effects of living around green things are powerful, but green is not the only color capable of soliciting measurable behavioral changes. Consider the color blue (470 nanometers and beyond). We've known for some time that blue hues keep brains alert and aroused. We even know why. Blue light suppresses production of the sleep-inducing hormone melatonin. The more blue you encounter, the more aroused you become and the more energy you acquire. The effect is so powerful, it can even interfere with sleep. Researchers recommend shutting off any electronic devices emitting blue light an hour or two before bedtime.

The arousing quality of blue may also have evolutionary roots. The color isn't something we'd normally encounter in the savannah's tawny grasslands, but our evolutionary history was hardly blue-less. To imbibe an initial dose of azure, we merely had to look up. Since we're not naturally nocturnal, it makes sense for blue to arouse us. In the mornings, it might function as an ancient African alarm clock, then something like Jolt Cola for the rest of the day. When night falls and the blue goes away, the energizing component of the wavelength fades as well.

And finally, light that encapsulates green, blue, and the rest of the spectrum (think colors of the rainbow ranging from 380 to 740 nanometers) is also good for you. Why? This spectrum is white light, or natural light, the kind you find when you step outside or look out a window. People who bathe themselves in natural light enjoy an 84% decrease in eyestrain, which includes reductions in blurred vision and headaches. People who are lucky enough to work in an office that immerses them in natural light don't take as many sick days either. Exposure to natural light also positively affects sleep patterns, including a longer sleep duration on average and a higher quality of rest. Natural light even affects retail space. Skylighted Walmarts have, on average, 40% higher sales than stores lit primarily with fluorescents.

Because the behavioral effects of natural light and the colors ' green and blue have Darwinian origins, interior designers would do well not to ignore them.

Give me a break (outside)

Given the strength of this happy marriage between health benefits and nature, how can we apply its advantages to the business world? Should regular breaks become a regular part of a worker's daily life, especially if the breaks can be taken in a garden? Healthy workers are more industrious workers, and healthy, *happy* workers are even more productive.

The simple truth is that people who take regular breaks outperform those who don't, and people who do it while basking in nature outperform everybody. This was first noticed in studies looking at on-the-job error rates. Put simply, people who took regular breaks made fewer slip-ups per unit time than those who did not. Employee focus improved too, as did task engagement, indicating upgrades in executive function (that important getting-things-done, cognitive gadget we discussed in the last chapter). It may feel like a contradiction, but people who spend less time on work statistically do their jobs more effectively than people who spend more time.

So, how often should you push the pause button during the day? It's here where the cognitive neurosciences provide real insight. Research shows you'll need to take a break every ninety minutes or so. If the job is excruciatingly intense, you'll need to increase that frequency. This figure comes from research by Nathaniel Kleitman. It was then championed by psychologist K. Anders Ericsson and businessman Tony Schwartz. Kleitman discovered something known as the BRAC, the basic rest-activity cycle. The brain, it appears, really wants a few moments to itself every hour and a half. Workers who take breaks according to that schedule increase their productivity the most, a fact demonstrated not only in the business world but also in the sciences and even the performing arts.

And what should you be doing every ninety minutes? Here's where you can cue E. O. Wilson. You should take a walk outside or at least visit rooms filled with plants and waterfalls. You'll remember I asked in an earlier section just how short a short walk in the woods has to be in order to impart any health benefits. Research shows that the tranquilizing effects begin quickly in such environs, within the first 200 milliseconds of exposure. Benefits increase the longer you sojourn in such places, anywhere from ten minutes to just under an hour.

I realize company policies, even state laws, are all over the map when it comes to taking breaks. If your company does have a break-taking policy, take advantage of it as much as you can, even if it feels counterproductive. And when you take that break, try to find some solace in nature by taking a walk outside or at least retreating to a place in the office with some foliage. If you are a policy maker at your company and don't have a policy on break-taking, make one. And as much as you can, try and get your employees to heed it. Most businesses run on the health of their employees' neurons. Underscoring the importance of keeping that human capital healthy was one of the reasons why I wrote this book.

Prospect-refuge

It seems clear that people who design office buildings should spend time looking at Japanese gardens. I'm guessing the poor people who designed the headquarters of SAS (Scandinavian Airlines) wished they had. Their 1987 creation consisted of a large central main-street floorplate, reminiscent of Disneyland. The floorplate connected various offices and conference rooms to a café, a sports area, and numerous open, informal meeting places. The idea was to give employees reasons to leave their stuffy offices and gather informally. The minor hope was that these encounters would allow cognitive elbow room for spontaneous, serendipitous interactions. The major hope was that they would lead to a boost in productivity.

Turns out, both hopes were a waste of time. An analysis of where employees actually interacted revealed that only 9% used the main street and its café. The other "open attractions" accounted for just 27% of the total. More than two-thirds of the employees still used their stuffy offices to meet with each other. So much for a boost in productivity.

Obviously, the designers had focused on the wrong things.

But where might they find the right ones? My vote goes in two directions: the wild, noisy slopes of Tanzania's Ngorongoro Crater and the less wild, less noisy academic halls of northeastern England's University of Hull. These places seem to hold the secret to some of the most creative, productive activity in the world.

The proposition that Tanzania and the United Kingdom might solve airline-headquarter design problems obviously requires some discussion. Let's start with something called *prospect-refuge theory*. The term was first used by University of Hull emeritus geography professor Jay Appleton. To understand Appleton's ideas, we need to describe what may have been his inspiration: the Ngorongoro Crater of eastern Africa.

The Ngorongoro is in Tanzania. The crater is actually a volcanic caldera, the largest inactive volcano in the world. It blew its top millions of years ago, creating a bowl-shaped valley almost 100 square miles in size. It abuts the Serengeti and is part of a larger complex called the Ngorongoro Conservation Area. Along the sides of the bowl-shaped caldera are uncomfortable, craggy ridges and steep hills, pouring onto flat-as-pancake plains. Many places along these ridges, especially in a region called the Engaruka, are pockmarked with caves. This geological feature gave hominids shelter and a quick place to hide. It simultaneously provided nearly measureless vistas where we could surveil predator and prey. In other words, it provided for us both prospect and refuge.

Because of the Ngorongoro's unique landscape and influence on our prehistoric ancestors, you could also call it *homo sapiens's* East African skunkworks. Our ancient brains were reinvented and modernized in regions like the Ngorongoro (and areas beyond), allowing us to develop into the current art-making, structure-building, observation-taking, tax-paying creatures we are today.

A bit more on prospect-refuge

What's clear from Appleton's ideas is that we prefer a mixture of environments. And since we spent 99.987% of our time on the planet in such an environment, it doesn't take the heft of E. O. Wilson to think Appleton might be onto something. What he's onto is a balance of two preferences.

On one side is prospect. We possess a species-wide need to be able to survey our surroundings with as large a vista as possible. Large views immediately tell us where water might pool, where predators might lurk, where prey might be lingering. Hunter-gatherer groups with a predilection for such knowledge will obviously survive better than those without. We still have those preferences, observable with something as contemporary as your neighborhood real estate agent. Houses with views sell, and people are willing to pay more for them.

On the other side is refuge. Appleton says we also prefer environments that allow us to hide from enemies, nasty weather, and, as our populations grow, each other. Being able to control access points— something caves allow—increases our chances of confronting threats directly. Being able to have a roof over our heads increases our chances of surviving the meteorological moodiness of even the Serengeti. One of the reasons that caves on the side of an ancient caldera could be so appealing is that they gave us both.

Appleton predicts that the most productive spaces in the world will have a balance between prospect and refuge. Empirical work seems to back up the claim. Two researchers in the *Harvard Business Review* agree:

> *The most effective spaces bring people together and*
> *remove barriers while also providing sufficient privacy*
> *that people don't fear being overheard or interrupted.*

This sensitivity to the expansiveness of our surroundings can even be observed by studying ceiling height, believe it or not. In what researchers call the *cathedral effect*, ceiling height affects the focus of professionals trying to problem-solve. The higher the ceiling, the more the subjects focus on the gist of the problem (least attention to detail). The lower the ceiling, the more the subjects focus on the details of the problem (least attention to gist). Implications? When professionals are solving large problems, they need to be in rooms the size of St. Patrick's sanctuary. When solving detailed problems, they need to be in a cave. Seems like even when we're sheltered, we still respond to prospects and refuges.

Design implications

This evolutionary need to work in the fulcrum between two spaces—open areas and enclosed shelters—provides insight about office planning. Since unbalanced designs disregard 50% of our prospect-refuge needs, you might predict that workspaces focusing on only one aspect—all open areas or all rabbit warrens—would fail.

That appears to be the case, and Exhibit A may be offices with open-concept floorplans. It's long been attractive to have these vast areas of inexpensive, wall-free cubicles accompanying exposed collaborative spaces. Spontaneous interruptions are encouraged, and serendipitous interactions are assured. But is that really the case?

Unfortunately, very few people tested this idea to see if such assurance was warranted. Turns out, when you do test it, the enthusiasm isn't warranted at all. Instead of discovering acres of happy, buzzing productivity, researchers find sullen, stressed battlefields, where the chief victim is the bottom line. Productivity plummets with open-concept office spaces. Creative thinking declines. The ability to concentrate on tasks deteriorates, and stress levels soar. Unsurprisingly, job satisfaction regresses when there's too much prospect.

One of the biggest stressors here is the inability to escape distractions, like being force-marched to listen to other people's phone calls. Researchers have given this annoying feature a name, "half-a-logue," so monikered because you can listen to only half the conversation. This is extremely distracting to our brains, and the effect isn't small. Errors in concentration (measured by visual tracking assays) rise a whopping 800% in the presence of half-a-logues when compared with controls.

This desperate need for the caves of the Ngorongoro Crater can be seen in workers' attempts to go back there, to make refuge when there's too much prospect. The *New York Times* reports it as something akin to a war:

> *The walls have come tumbling down in offices every-*
> *where, but the cubicle dwellers keep putting up new ones.*
> *They barricade themselves behind file cabinets. They*
> *fortify their partitions with towers*
> *of books and papers …*

Headphones can help, though they're hardly an ideal solution. We still respond to distracting visual cues even with noise-cancelers firmly covering our ears, which is an example of the need for Uncle Darwin to come to the rescue.

The point here isn't to *eliminate* open-concept office areas. The point here is to *balance* them. Done properly, serendipitous interactions can boost productivity by 25%. We are, after all, a naturally collaborative species. That said, we are social because we're interested not in chatting but in surviving. This priority requires us to hedge all the socializing with time for ourselves. If it's all open space all the time, it's too much of a good thing.

What to do next Monday

Since this chapter is stuffed with practical ideas, I think it would be wise to apply some of its content to a thought experiment: If you had an unlimited budget, few bureaucratic constraints, and could design work environments based on the topics in this chapter, what would they look like? Now is a lovely time to engage in such an experiment, especially as we grow out of our post-pandemic existence. The very notion of an office has become the subject of much speculation and debate, from interactions within a home to the necessity of a building. It's likely that the concept of an office edifice is here to stay. But during this season, when ideas are still malleable, why not go all the way and reimagine everything?

Let's begin with the office building and, more importantly, Jay Appleton. You'd plan the edifice with his prospect-refuge ideas as the centerpiece design element, I'd warrant. There would be spaces for experiencing large, idea-generating vistas. These spaces would be physically connected to private offices, where employees generating those ideas had quiet spaces to develop them. Sound far-fetched? We already have formal, competent design devices for prospect-refuge. They're called balconies.

And what would those balconies overlook? An arboretum. The ideal building would be marinated in green space, either internal spaces overlooking garden-like courtyards or, more practically, external outdoor spaces suitable for forest bathing. Employees would have regular access to trails, streams, and waterfalls, and be able to see these features from their offices.

Our fantasy also incorporates floorplate design. Offices would be filled with natural light, every space rife with immunity-boosting plants (a money-saving idea, especially during the winter months of flu season). Employees would be allowed to visit the botanical spaces to take regular breaks every ninety minutes, preserving valuable human capital.

Even the conference rooms would be unique, rebadged as "problem-solving rooms." Lighting would be adjustable in these spaces: green when workers need to focus, blue when they need energy. The height of their ceilings would be modifiable too, cranked high for problems requiring 40,000-foot solutions, cranked low for problems requiring detail.

These ideas have revenue-friendly benefits. And unlike some of the more unfortunate and unproductive office designs of the past, they also have the support of scientific study baked into them.

I realize that turning an idea into a reality isn't the easiest (or cheapest) thing in the world. Yet, as a brain scientist, I also realize offices weren't designed with the brain in mind, though nearly every business on this planet relies on the organ to keep its bottom line in shades of black. Most companies haven't even tried to consult the cognitive neurosciences, though we've had more than enough time. (Some of the data presented in this chapter are *decades* old.) While some of these notions are as expensive as Christmas, some are as affordable as a spider plant. All of them sit on a pile of peer-reviewed papers, atop which I can imagine E. O. Wilson gleefully dancing.

Thousands of office workers, perhaps millions, would love to be able to join him.

THE BUSINESS OFFICE

Brain Rule: *The brain developed in the great outdoors. The organ still thinks it lives there.*

- Humans have spent 99.987% of their history living in natural environments, and modern living can push us into prolonged periods of stress called *role overload*, which can cause burnout, mental fatigue, and even brain damage if left unchecked.
- To reverse the processes that induce stress, find ways to get yourself (and your subordinates) exposed to natural elements.
- If the great outdoors aren't readily accessible at your location, incorporate elements such as natural lighting, an abundance of plants, and the colors green and blue into the office design.
- Take a break, if you can, every ninety minutes—even better if you can take a break outside.
- To foster employee creativity, design a workspace where both prospects (open public areas) and refuges (enclosed private areas) are available to all.

creativity

Brain Rule:
*Failure should be an option—
as long as you learn from it.*

I'M GOING TO BEGIN THIS CHAPTER by asking you to think of as many new uses as you can for a brick. I'd like you to make a list of the things you come up with. I'll wait. I'm in no hurry. I wouldn't want to look at your answers until you get to the end of this section anyway.

The reason I'm asking for this list concerns this chapter's subject: creativity. We're going to talk about how to define creativity (hard to do), what hurts it (harder to do), and finally how to boost it (hardest of all to do).

Brain scientists have historically had a hard time studying creativity—not because we don't believe it to be sufficiently real, but because we don't believe it to be sufficiently quantifiable or characterizable, at least with current technologies. We're not sure what we're looking at or looking *for*.

Most of us would agree that Leonardo da Vinci was creative. Most would say Einstein too, and Beethoven and George Balanchine and Louis Armstrong. Were their brains all drinking from the same creative well while creating their masterworks? Were they accessing the same brain regions to come up with their famous outputs? We have no idea. What we're left with is making lists.

Speaking of which, how did yours turn out? Did your list include uses like "paperweight" or maybe "doorstop"? Did you think to put your brick on the lid of a pot of boiling water to keep the liquid from running over? These are fairly standard answers to such questions. What unites them is that they don't deviate much from the weight-bearing properties of "brick-ness."

Some researchers would rank these answers lower due to the fact that they don't diverge much from what a brick was originally designed to do. Here's an answer that would score high though: pound the brick into dust, then use the powder to color paint. This new use diverges greatly from a brick's original function. Researchers put a lot of stock in this divergence. In fact, these stretches are what they're measuring when quantifying certain types of creativity. Such tests are called *divergent thinking* assessments.

Would you call one answer more creative than the other? Most people would nominate the paint pigment as more innovative—I certainly would—and many scientists would agree.

Divergent thinking is just one of several types of creativity researchers have attempted to characterize. We will take a look at several others. Along the way, we're going to discover if there are ways to turn someone's doorstop into a jar of paint.

Novel or nonsensical

Do you think the sentence written below is creative, or do you think it's crazy?

> *I am here from a foreign university ... and you have to have a plausity of all acts of amendment to go through for the children's code ... and it is no mental disturbance or putenance ... it is an amorition law.*

This is an actual speech, uttered by a real person, taken from a research paper published in the middle of the last century. From one perspective, the creativity is off the charts: the speaker starts making up his own words. New uses for a brick, vocabulary edition. This is nice while it lasts, I suppose, but does the sentence make any sense to you? It takes innovative liberties with the English language, but is there any utility in it? The statement is actually difficult, perhaps impossible, to understand.

So again, I put it to you: Is the sentence creative or crazy?

This is just one of the many dilemmas researchers face when trying to define creativity. What's the difference between the novel and the nonsensical? What behavioral scalpel can researchers use to tease out the silly from the sublime?

No such scalpel exists, sadly. Over the years, courageous souls have put forth ideas on this subject, most of which have proved humbling. The definition most researchers use springs from one such brave endeavor:

> ... *the production of an idea or product that is both novel and useful is commonly accepted as a central characteristic of creativity.*

"Commonly accepted" is science-speak for "we are going to punt on this one." It is the scientific equivalent of surrender.

Though the definition is far from complete, it does offer some utility. The distillation of creativity to two conceptual anchors, novelty and usefulness, has created a few testable ideas and some interesting scientific insights into the nature of innovative thinking.

One of the most valuable of these insights was a result of evolutionary biologists trying to understand creativity's original Darwinian utility. Creativity, the biologists concur, most likely derived from the need to navigate ancient climate change. Turns out, the climate in Africa during the last few hundred thousand years of our Serengeti sojourn was very unstable, shifting from hot and humid to cold and dry—sometimes within a couple of generations. This instability brought all kinds of new challenges to our fragile hunter-gatherer survival. Those who could apply brand-new solutions to brand-new problems adapted best to the change. Those who weren't innovative enough perished. Creativity conferred a survival advantage because it allowed our species to roll with the meteorological punches.

Given this evolutionary origin story, the utilitarian, second anchor of our definition of creativity is self-evident: The innovative solution must have some type of *functionality* to be evolutionarily advantageous. That's why there are two anchors and not one.

A "plausity of all acts of amendment" may sound creative, but it's not enough to save us from environmental instability.

Just so you know, the man responsible for the quote at the beginning of this section suffered from schizophrenia. Researchers give these tumbling, jumbling verbals a name: *word salad*. It's a common symptom of certain forms of schizophrenia. The words of this patient were certainly unique, but by the definition we are using, they were not very creative.

Convergent versus divergent

It's one thing to define creativity as needing to be both unique *and* utilitarian, but it's another thing to identify the discrete neural substrates—if there are any—that undergird these twin characteristics. Scientists usually address large questions like these by proposing models, testing them, then looking for relevant regions in the brain that might explain how the models work.

We will investigate three models in this chapter: divergent/convergent thinking, cognitive disinhibition, and an old phenomenon simply called *flow*. All are chock-full of testable ideas. All have been examined by experts whose job is to map outer-behavioral function to the watery inner world of the brain.

Remember when I asked you to think of new uses for a brick? That's an exercise in divergent thinking. This is the cognitive gadget that challenges a person to conjure up as many innovative ideas as possible—given certain parameters—in a very open-ended, unpressured way.

The second model, convergent thinking, is practically divergent's mirror opposite. This cognitive gadget challenges a person to conjure up many unique, creative solutions to solve a single problem. The solutions must *converge* on the task.

A master-class example in convergent thinking is observable in *Apollo 13*, the movie based on the true story of how NASA dealt with a famously crippled spaceship of the same name. All kinds of innovative problem-solving were depicted in the film (one example involved a training manual cover, a bungee cord, and socks!). The singular goal was to turn the ship around and bring the astronauts home alive.

Since the difference between these two types of creativity can be confusing, here's a simple way to tell them apart: Think of divergent thinking like a firecracker, with multiple multicolored arcs exploding from a central point. Think of convergent thinking like a magnifying glass, focusing numerous points of light onto a single source.

Ever mindful of practical implications, researchers naturally ask what kinds of phenomena help and hinder divergent/convergent thinking processes. Stress turns out to be a major player in both, but it affects the two brands of creativity in very different ways. On the one hand, stress can be a powerful and motivating friend to creativity, especially the convergent-thinking variety. It was a real incentive for the engineers at NASA to think outside the box when the three lives inside the Apollo 13 space capsule were at stake.

Yet certain types of creativity, such as divergent thinking, fold like wilted flowers when stressful conditions are encountered. When people feel rushed or pressured, they don't score very well on tests that measure divergent thinking. (That's why we talked about not being in a hurry in the introduction to this chapter.)

We can see statistical proof of creativity's codependent relation-ship with stress. One of the greatest predictors of long-term creative output in people is how they handle failure. For some people, failure is an extremely stressful experience. For these people, the prospect of failure stifles any otherwise innovative instinct they might have had. For others, failure is no failure at all; it's just another helper, assisting courageous innovators on the way to the right solution.

Failure, never failing

Researchers from a broad variety of disciplines have explored the uncomfortable link between creative output and the fear of failure.

Just look at the titles of the papers being published. From the business side, you get headings like "The No. 1 Enemy of Creativity: Fear of Failure." From the neuroscience side you get titles like "Fear Shrinks Your Brain and Makes You Less Creative." That shrinkage occurs in several brain regions, by the way, most notably in the hippocampus. That's a big deal. The hippocampus is involved in many processes important to innovative output, including transforming short-term memory traces into long-term forms. Shrink the hippocampus enough, and you interfere greatly with this processing function.

Why should fear of failure do such negative things? And why is our innovative instinct such a special target? To answer these questions, we have to discuss something all babies, scientists, and entrepreneurs have in common. It concerns how they learn.

Babies learn through a series of increasingly self-corrected ideas, using hypothesis-testing software pre-loaded at birth. They are (a) constantly making observations about how they think their world works, (b) testing their ideas in trial-and-error fashion, and (c) modifying their understanding based on the data obtained. If you think that sounds like something scientists do—good old scientific methodology—you're right on the money.

Many years ago, a book entitled *The Scientist in the Crib* showed just how extraordinarily similar babies and scientists are (and I can tell you from personal experience, there is more than one point of intersection). Such hypothesis-testing styles are potent. Iterative, repetitive processes are muscular enough to fling rocket ships to distant asteroids and gentle enough to ferret out the secrets of the atom.

These processes are also littered with failure. In fact, the notion of failure is cooked into the mechanism; it's the *error* side of the trial-and-error equation.

You can see examples of this link between failure and creativity everywhere. Very few entrepreneurial projects succeed the first, second, or even twentieth time they're attempted. An astounding number of theories posed by scientists fail under rigorous testing. Even successful scientific theories rarely come out the other side of testing unmodified. And I don't know of a single baby that didn't spent wobbly weeks—sometimes months—stumbling around, rising up, and falling down before they could stand up and move forward.

If you're paralyzed by failure, the paralysis will extend to your project and ultimately to your productivity. So getting your attitude toward failure right is really important. Remember that one part of creativity is coming up with novel ideas, while the other is making them useful. Failure, it turns out, is the engine behind turning something fantastic into something functional.

Maximize your failure's potential

There's business relevance to these ideas. Let's revisit Google and their Project Aristotle (the previously discussed project of investigating why certain teams work so well). The researchers concluded that psychological safety was the key to success, allowing interpersonal risk-taking to flourish, which, of course, includes accepting failure.

Substantial progress about the connection between psychological safety and innovation has been made since Project Aristotle's time, including the ability to quantify risk-taking norms. One experiment found that groups who (a) tested multiple ideas simultaneously (between three and five proposals) and then (b) selected the top two or three results for further experimentation achieved a 50% higher success rate than those who didn't iterate or didn't iterate enough. Losses, like cookies, can be baked in batches.

Perhaps inspired by these efforts, more researchers began cooking in failure's test kitchen. Soon it was discovered that permission to fail wasn't by itself predictive of victory. People who achieved great success and people who suffered spectacular failure tried and failed roughly the same number of times. The difference? The folks who succeeded tried to learn something from their errors. They faced their errors squarely, squeezing their efforts out like dishrags.

Those who didn't have such occupational courage and tenacity were more likely to continue failing.

Another finding concerned the amount of time people permitted to elapse between consecutive failures. The less time they allowed, the greater their chance of future success became. The longer people tarried between efforts, the more likely they were to continue to fail. It was not only good to learn from failures but also important to get back on the horse and try again as soon as possible. Such rapidity can only occur if people feel free to fail.

One company known for allowing its employees to fail was IBM, helmed under the secure leadership of legendary CEO Thomas Watson, Jr., He headed the company during some of its most successful and innovative years. In one famous story, an IBM VP once tried an experiment that failed and cost the company almost $10 million. His mea culpa was a resignation letter, which he delivered to Watson personally. The VP was stunned by his boss's subsequent reaction. "Why would we want to lose you?" Watson said with a laugh after reading the letter. "We've just given you a $10 million education."

It's true that businesses are not charities; Watson wasn't out to break IBM financially, certainly. But perhaps he knew intuitively what, these days, you can show empirically: you increase your chance of winning by increasing your tolerance for losing.

Tech columnist Michael S. Malone puts it like this:

> *Outsiders think of Silicon Valley as a success, but it is, in truth, a graveyard. Failure is Silicon Valley's greatest strength."*

Breaking the shackles of fear

So, what should you do if fear-failing shackles you to innovation-failing? Are there practical attitudes you could cultivate to loosen this toxic linkage? The answer is yes. It begins by understanding why we fear failure.

Research shows that many employees view their errors as a personality flaw. For those who feel this way, failure isn't merely something they did wrong; it is a referendum on who they are as a person. If you feel that way, you may try to hide your mistakes. Lie about them. Blame somebody else for them.

People who do not see their flops as flaws—something to be ashamed of, lie about, or offshore to somebody else—actively run toward their disappointments. And they reap the rewards of their courage, becoming more successful in the difficult trial-and-error world of creating the next best thing. Indeed, there is evidence that failure acts as an accelerant to most people's red-hot ideas. As psychologist Robert Epstein couched it, "... *failure actually stimulates creativity directly. It is really valuable.*"

This suggests you can cultivate a healthy attitude toward failure by following a simple three-step protocol. This protocol is best illustrated by, oddly enough, a Florida fireman. He's Matt Holladay, a strapping, muscular man who shaves his head and looks like he's from first-responder central casting.

Holladay was training a group of new recruits one bright Florida day when an emergency call rang out—a residential fire, with the house mostly in flames. With colleagues in tow, he rushed to the house, stopped, and assessed the situation. Smoke was billowing from the entire house except for one bedroom. There was a chance someone might've been inside. He immediately jumped through what had been a window and landed right beside a frail old grandmother. She was still alive! He picked her up, passed her through the window opening to his waiting colleagues, then quickly got out himself.

Holladay's action story is divisible into three steps: First, he ran toward, not away from, the inferno. Then he assessed the damage, looking for possible signs of life. Finding a way, he did something about it. He entered the bedroom, discovered he was right, and saved someone's grandmother.

Believe it or not, research says to follow a similar three-step protocol when responding to a failure.

Here are the three steps:

1. *Run toward the failure as if you were Holladay running toward a fire. Research shows—as does logic—that willingness to engage a threat is the single most important factor in neutralizing it.*

2. *Begin assessing the situation. Determine whether a grandma is in the room and then figure out a way to rescue her, even if everything else is burning down around you.*

3. *Learn as much as you can from your assessment. Find out why the house was burning in the first place, then take action steps to address what went wrong.*

Turning Edsels into Mustangs

There are many business examples showing how these three steps lead to productive outcomes. One famous illustration comes from noted business guru Peter Drucker, who famously described the story of one of Ford Motor Company's biggest flops: the 1958 Edsel. The car was overplanned, over-researched, and spectacularly over-sold. When the car was presented to the public, sales were abysmal. *Business Insider* estimates the Edsel cost the company $350 million.

Drucker related that rather than recoiling from this expensive dumpster fire, the executives in charge ran toward it. In a deliberate, systematic fashion, they investigated what failed, discovered what succeeded, and ascertained ways to turn one into the other. Their efforts led directly to modifications that resulted in the creation of the Thunderbird and the Mustang, two of the most successful car models of all time.

So, what are the conditions that allow companies to turn Edsels into Mustangs? Drawn from decades of research, the answer has proven to be consistent, robust, and, if you're an executive, slightly unnerving. It's how executives react to failure—their own and that of their colleagues—that directly affects how everyone else reacts too. The through line to productivity is contingent not on an action but on an *attitude*.

Like I said, unnerving.

If you are an executive or manager, what attitude should you cultivate when it comes to failure? In order to turn disappointment into dollars, you have to create an atmosphere where failure is not only accepted but also expected. And it all starts with leading by example, something behaviorists call *passive transfer*. Managers who consistently put on their first-responder clothing whenever failure occurs create some of the most productive companies in the world.

An executive named Ed Catmull once said, "Mistakes aren't a necessary evil. They aren't evil at all. They are an inevitable consequence of doing something new." Catmull should know. He was the cofounder of Pixar.

Innovation climate control

How do you create an atmosphere where it's okay to fail? Research suggests there are short-term and long-term solutions.

Short-term solutions involve addressing your own psychological interior—managing your reaction to failure, particularly how you deal with fear. If you, as an executive, are afraid of failure, that fear will percolate to others. Fear, as every experienced first responder knows, is contagious.

Often, managing our reactions to fear is a matter of impulse control. It's natural, after all, to be reluctant to enter a burning house. But there's also hope here, for research shows how to deal with impulse control. The behavior is a card-carrying member of Team Executive Function, the cognitive gadget described previously as "the ability to get things done." Anything that improves executive function is going to improve impulse control.

Mindfulness, a meditative protocol, improves impulse control. Physical exercise does too. They begin exerting their effects in a surprisingly short period of time—less than a year—which is why I call them short-term solutions. Not surprisingly, mindfulness and exercise both improve creativity, and in specific ways. Focusing the mind improves divergent-thinking scores. Exercise does too, especially if you can do it outdoors. Turns out, Henry David Thoreau, the original forest bather, was right when he said, "The moment my legs begin to move my thoughts begin to flow …"

The long-term solution is more process-oriented. Research suggests that companies should create deliberate bureaucratic gadgetry to both generate and develop in-network innovations. These mechanisms should be written down and given to everyone who's ever had a good idea. The crown jewel in this document? An explicit exposition of how to deal with failure. Here's a quote from researcher Kevin Desouza et al.:

> *The optimum situation is for management to have*
> *a well-defined innovation development process that*
> *includes failures as an accepted part of business.*

Desouza et al. even know what that failure-friendly process should look like. They describe a total of five features, which can be thought of as steps in a cycle. The first three steps are especially important for our purposes:

First, there is an "idea generation and mobilization" feature. This step involves creating an open-ended environment, friendly for idea incubation, and an immediately accessible log to make people aware of the efforts.

Second, a deliberate vetting process needs to be installed, one where the pros and cons of any idea can be evaluated. (They call this a "screening and advocacy" step.)

Third, a process/mechanism needs to be built that allows people to test their ideas and, where appropriate, begin building prototypes. These efforts may involve labs, or at least places where tinkering occurs.

The last two steps involve commercialization and getting buy-in from potential consumers. Marinated throughout all these steps is the explicit acknowledgement that it's okay to be disappointed. I have a sign that should be put up in every room where this process takes place, another quote from Pixar's Catmull:

> *If you aren't experiencing failure, then you are making*
> *a far worse mistake: you are being driven by the desire*
> *to avoid it.*

Words to live by, especially if you want to make billions of dollars, which Ed has actually done.

The second model: cognitive disinhibition

Sadly, research shows that most companies don't have anything formal like this five-step cycle in place. They simply rely on random serendipity, whatever *that* is, and also on their own instincts—which, if based on fear, is mostly useless and to their peril.

Research shows what logic implies: Companies that are the most innovative are the ones most likely to lead. Having a formal mechanism to try things out in a failure-friendly environment is necessary but may not be sufficient.

Divergent/convergent models aren't the only theories of creativity out there and may not be the only models companies need to succeed. Our next model of creativity may help fill the gap. It consists of an opaque concept called *cognitive disinhibition*, whose complex contribution to creativity consists of dual parts. Fortunately, both can be explained by a simple analogy from a particular scene in the Broadway musical *West Side Story*.

No kidding.

The movie version of this scene involves a fateful moment when two star-crossed lovers, associated with rival 1950s-era gang members, meet and fall in love. The moment occurs in a school gym, where a dance is being held as a reconciliation event for the gangs—a space where they might solve their conflicts after partying together.

Reconciliation, sadly, is not in the cards for them. The event devolves from a dance to a near-riot, confused with noisy clashes of music and the choreographed collisions of dozens of bodies. The increasingly disorganized cacophony reaches a climax when, suddenly, two lovers, Tony and Maria, spy each other from across the floor. As their eyes lock, the music fades to silence, and the camera puts everyone else out of focus. They dance politely—nothing else and no one else in the gym.

This scene always reminds me of the two contributions cognitive disinhibition makes to creativity, which won't make sense until we define it. Here's a quote from researcher Shelley Carson:

> [Cognitive disinhibition is] the failure to ignore informa-
> tion that is irrelevant to a current goal ... a mental filter
> that willingly chooses not to block information coming
> into conscious awareness.

Carson's definition is the equivalent of allowing noisy dancers to run wild in your brain. As you problem-solve, you give free rein to large numbers of inputs. None of these are necessarily meaningful, but they are still allowed to roam freely on your cognitive dance floor. You are *disinhibiting* your conscious awareness, hence the term.

If these disinhibiting dances were the only things happening in your brain, you wouldn't satisfy our working definition of creativity. As we discussed, you could responsibly say the patient with schizophrenia—the one advocating a "plausity of all acts of amendment"— was very disinhibited, but his disinhibition went no further. Nothing meaningful or practical came from his speech. Happily, even poetically, the missing ingredient is demonstrated by Tony and Maria.

Removing distractions

The *West Side Story* gym scene involves highlighting a subset of inputs (a budding romance) by removing distracting information (an unruly dance).

As a creative person watches multiple inputs pirouette around their conscious awareness, they may notice some of those inputs are starting to gaze at each other. A relationship may be forming, perhaps based on something in common. Or perhaps nothing in common. There may be properties of these inputs that generate something useful when put together in a certain combination. Whatever the reason, they're not like the other dancers.

As soon as that recognition occurs, truly creative people can willfully fade out all the other distractions. They then can concentrate only on the exciting inputs. Researchers call this behavior the ability to focus/defocus attention. From this focusing/defocusing, creative insight springs. Utility often follows.

These two abilities—a generous willingness to entertain a cognitive rave followed by a ruthless ability to ignore most of it—lie at the heart of cognitive disinhibition. The difference between crazy and creative might very well lie in the ability to discover star-crossed ideas gazing at each other from across the room, then allowing them to dance.

The brains behind it all

The behavioral evidence in favor of cognitive disinhibition as an actual thing is strong. For that matter, so is the evidence for divergent/convergent thinking. What's less clear are the neural substrates underlying these behaviors. Researchers spend a great deal of effort trying to find the innovation regions of the brain. Sadly, we're still looking.

We know some of the neural ingredients that must be activated for the outwardly observed behavior to be internally observable. In the case of cognitive disinhibition, we know that working memory is critically involved.

Working memory is a cognitive space where large numbers of inputs can be temporarily stored. It used to be called *short-term memory*. This volatile buffer is neurologically underwritten primarily by brain regions that exist right behind your forehead (prefrontal cortex).

How does working memory relate to cognitive disinhibition? It comes from the need for a cognitive space that can hold many things at once, some buffer where disparate inputs can interact in real time. That buffer is what working memory provides. It's your brain's equivalent of the gym in *West Side Story*. You won't have much of a dance without it.

No need to yell

As you might suspect, the size of this mental gym directly influences creativity, and for a simple reason: anything affecting its volume will also affect the number of variables it can hold.

So, what types of experiences affect the size of the gym? You probably already know the answer, given our already long workout with this concept. Loss of control—negative stress—affects the moment-by-moment capacity of working memory.

There are many ways to show this. One real-world set of experiments investigated what happens to people's working memories when someone becomes verbally aggressive. *Verbal aggression* is the scientific euphemism for *yelling*, something many subordinates experience from cranky bosses.

For the sake of argument, suppose you're the cranky boss. You decide to let your emotional clutch slip after observing an error, and you begin yelling at the responsible employee. What happens to your employee? Your actions immediately shrink their working memory by a whopping 52%, which deeply disrupts the memory buffer's carrying capacity. This shrinkage can affect their creative output in virtually every way we can measure creative output.

Further research provides a hint as to why such memory shrinkage occurs. The hint comes from an unexpected place: law enforcement and eyewitness testimony.

Mental health professionals know that trauma and memory loss affect each other in a reciprocal fashion. When something bad happens to a person—say, an assault—they usually experience some level of amnesia, especially for events around the immediate time of the assault. This can directly affect their eyewitness testimony.

The amnesia isn't usually total, however. If the trauma involves a weapon of some kind, such as an assault with a gun, things change considerably. The brain's memory systems will push the recording button to remember every detail it can about the firearm. Severe memory shrinkage still occurs, but it might be better to call it a shift, a hyper-reallocation of resources. Such abnormal concentration at the expense of just about everything else is called *weapon focus*.

This phenomenon directly relates to our yelling story. If you become verbally aggressive toward someone, you have essentially turned your mouth into a weapon. Your subordinate's tracking radar will automatically target the source of the threat, which is *you*, while their other memory systems fall away. Rather than focusing on what may be a legitimate concern—like the error—the subordinate focuses on the opposite, your angry mouth.

Some managers ignore this warning, thinking verbal aggression increases innovative productivity. It doesn't. Yelling no more increases creativity than brandishing a gun calms people down.

On the matter of flow

You get ten points for creative problem-solving if you know how to pronounce the last name of this famous Hungarian-born scientist: Dr. Mihaly Csikszentmihalyi. He's the author of our third and final model of creativity, a model he terms *flow*.

115

Give up? Csikszentmihalyi is pronounced *cheeks-sent-me-higher*. His ideas about productive creativity are as unusual as his surname.

The central idea of the flow model concentrates not on a person's output but on the internal condition of a person caught in the act of creation. In a flow state, a person hyperfocuses on some endeavor with such potency that it changes their psychological interior. They lose perception of time. They quit listening to the distractions of their thinking minds. They eventually become intoxicated with the process. Soon, little else seems to matter. In fact, the joy of creation can be so all-consuming, people will strive to continue flowing even if there's a cost for doing so. The result is a sense of well-being. In perhaps the ultimate triumph of pure curiosity, the activity becomes an end unto itself.

Csikszentmihalyi believes the flow state is one of the most pleasurable feelings a human being can experience. He also believes that flow isn't something that just happens when people seek to become creative. It isn't automatic. Rather, flow occurs when a set of conditions are met, like a finicky plant requiring specific nutriments in the soil.

One of the biggest constituents of this "soil" involves selecting a task with this specific Goldilocks characteristic: challenging enough to keep you interested but not so overwhelming it becomes impossible to achieve. Obviously, selecting tasks that fit into your skill set, even if that fit is uncomfortable, is key.

Goldilocks isn't enough, however. The "soil" also must contain clearly defined objectives within the selected task. Along with that, you'll need to create an immediate internal feedback mechanism, telling you if and when the objectives are achieved.

The final ingredient focuses on "where-has-the-time-gone?" issues. Your ability to focus on what's happening in the moment—even enjoy its immediacy in a nonjudgmental way—is a key soil component. If you think that sounds like mindfulness training, you're right on the money. Focusing on the here-and-now, as opposed to the there-and-later, allows action and awareness to fuse. You become immersed in whatever you're doing, permitting the flow to grow into a time-compressing blossom capable of making hours feel like seconds.

Networking

No one model accounts for everything we know about innovation, not even the three specimens I just described. And things just get more confusing when one tries to identify the brain regions behind the behaviors these models measure.

Many research dollars have been spent exploring the nerves behind divergent thinking—that new-uses-for-a-brick task. It's been remarkably slow going. One difficulty has been designing specific behavioral tasks capable of measuring divergent thinking *only*, uncontaminated by anything else.

Happily, progress hasn't been zero. There's evidence that a trio of neural networks team up to help you turn a brick from an anodyne doorstop into new forms of paint. These networks should sound very familiar:

The first is the DMN, the default mode network, the one involved in daydreaming. Not surprisingly, it's hypothesized to be the primary generator of creative ideas.

The second concerns the EF (executive function) networks, the ones mostly responsible for us getting things done. You can see the two pillars of our definition of creativity at work in the brain here: novel notions that are also useful. DMN gives you the ideas, and the EF puts them to work.

The third is the SN (salience network), the system usually involved in threat detection and response. These nerves may have a side hustle, examining the daydreams, then making value judgments about what information is worthy enough to pass along to the executive branch.

This triple interaction is hardly a one-size-fits-all explanation of divergent thinking or any other type of creativity. The greatest utility of this Holy Trinity Network hypothesis is that it provides testable ideas.

Researchers have also attempted to characterize the biochemicals behind creativity. Scientists have been studying the aha moment of creativity, for example, the experience when things come together and some new insight emerges. This work is done using a psychometric instrument called the Remote Associates Test. You're given three words, then asked to find a fourth term that would work to form a relevant compound word from the others. If I gave you the words *aid*, *rubber*, and *wagon*, for example, you could respond with the word *band*.

This test is conducted while you're sitting in a brain-imaging machine. Researchers discovered that the instant you land on *band,* the brain lights up like a blowtorch and in very specific patterns. One pattern includes areas that manufacture dopamine, the chemical responsible for reward and pleasure. It's absolutely delightful to think that your brain gives you a dopamine lollipop every time you figure something out. This has been happening to you since the first grade.

The power of Earl Grey

Unfortunately, you're no longer in first grade. Adult bodies need rest, regardless of age, and when that's not possible, they need pick-me-ups. You spend most of your workdays ping-ponging between the need to yawn and the desire for a double tall nonfat latte.

One of the most startling findings about rest is how profoundly it influences creative output, specifically problem-solving abilities. The literature is littered with papers with titles like "After Being Challenged by a Video Game Problem, Sleep Increases the Chance to Solve It."

One delightful research paper of this ilk involved giving subjects a series of puzzles to solve, varying when the subjects slept as they tried to solve them. Overall, subjects that were allowed an eight-hour period of sleep in the midst of solving the puzzles had triple the success rate of the controls. Researchers even know when problems should be consciously worked on. It's best to reexpose yourself to unsolved problems just before bedtime.

Why is this the case? Years ago, it was discovered that your brain doesn't go to sleep just because you do. The organ simply changes functions. It activates "offline processing" systems, repeating what you've learned during the day, working on creative ways to solve problems you encountered earlier. The REM stage of sleep—in which your eyes are moving (REM stands for rapid eye movement)— is used for creativity. The non-REM stage of sleep—in which your eyes aren't moving—is used for memory consolidation.

In stress-filled workplaces, it isn't always possible to get a good night's sleep, of course. But even here, brain science can be useful, exploiting one of the most abused psychoactive substances on earth: caffeine.

Caffeine consumption enhances creativity in a startling number of ways. It potentiates working memory, sharpens focusing ability— both signature behaviors of cognitive disinhibition—and increases the energy available for both. Caffeine also boosts convergent and divergent thinking scores. The delivery system for caffeine doesn't matter, although the effects on types of creativity do. Certain types of tea enhance *divergent* thinking abilities, for example. Almost any kind of coffee will boost *convergent* thinking scores.

Interestingly, caffeine doesn't work by directly stimulating your nervous system. It works by blocking your ability to feel tired. (For the biochemists in the room, it thwarts the ability of adenosine to bind to the brain's A1 receptors.) So you continue to expend energy—actually, *over*-expend energy—under caffeine's direction. When the drug wears off, you become doubly tired. Which simply means you need to sleep it off, which then becomes another way to enhance creativity. Like I said, ping-pong.

Aging and shifting

I've spent most of my academic lecturing career teaching graduate students and post-doctoral fellows. From a generational perspective, they're really a fun group to teach, students ranging from "kids" in their twenties to "kids" in their thirties and forties. When I lecture about creativity, I'm not surprised when I'm asked an age-related question: "At what age are we our most creative?"

I tell them we actually know the answer to that question: Expected creativity at age forty (most people's peak) is about double what it is at age eighty. But I flag this answer with some really important caveats. Some people make one great creative contribution in life, then never make another. Others remain ridiculously creative their entire lives. (Frank Lloyd Wright designed one of his most iconic works, the Guggenheim Museum, when he was in his nineties!) Also, there's evidence that older people are wiser because of their greater fund of knowledge and experience. This knowledge—unavailable to twenty-, thirty-, and forty-year-olds—can inform the creativity of those younger groups as they accelerate into their peak years. *But only if the generations are allowed to interact.*

I then launch into a discussion about the work of the previously mentioned psychologist Robert Epstein.

Epstein, who's roughly my age, invented what's called the "shifting game." He uses it to teach people the power of team-based creativity and, interestingly, solitary contemplation. He typically divides game participants into two groups, then asks them to solve a divergent thinking problem.

The first group, the control, is allowed to deliberate for fifteen minutes, listing as many uses as possible. The second group, the "shift group," is allowed to deliberate for only five minutes. Then they're told to leave the room, find some solitary place to think, and continue working on the problem. After five minutes, these individuals are asked to reassemble and create their lists of potential solutions. Epstein found that the shift group usually produces twice as many solutions as the non-shift group.

"And given what I just said about age-related creativity," I say to my students, "who should populate Epstein's teams?"

Epstein found that to make a shift group as creative as possible, given these confounders, we should make it as intergenerational as possible.

I usually conclude this part of the lecture by saying a variation of, "Just look around this room. You are right now in the midst of one of the most creative groups on earth."

Then I ask them to go find new uses for a brick.

What to do next Monday

In the 1970s, a couple of Seattle teenagers got an idea for improving car-counting on roadways. They used primitive computer-based technology for their solution, then formed a company around their product. When they first demonstrated their solution to the powers that be, their efforts failed spectacularly (the machine wouldn't work). Undeterred, the teens tried again and had a bit of modest success, though nothing to write home about. They eventually left the project, motored off to college, but continued seeing each other. The real value wasn't in the company, it turns out, but in those continued interactions, fueled by healthy reactions to failure. As we'll see, these attitudes and accompanying interactions changed the world.

Those reactions lie at the heart of our discussion of creativity, which we now dress up in practical overalls. Though we covered a broad range of topics, we can distill the next-Monday deliverables into seven suggestions:

1. *Decide which type of creativity is required for a task.*

Projects requiring divergent thinking will need different approaches than projects requiring convergent thinking. To be effective, divergent thinking requires freedom from things such as a time restraint, a stressful atmosphere, and fear of failure. Convergent thinking can actually thrive in stressful environments.

2. *Learn to run toward—rather than from—failure.*

This means investigating your failures, then summoning the personal courage to learn from them. When people observe that you view failures as learning exercises, they will follow. Leadership behavior, as we'll see in the next chapter, is highly contagious.

3. *Normalize iteration.*

Get ready to make three to five solutions to a given problem, then test-drive them. Don't let much time pass between failures. Show everyone that Silicon Valley's most visited monuments are its entrepreneurial graveyards. This is much easier to do if you've followed suggestion #2.

4. *Foster a climate of security.*

Ensure that your colleagues feel psychologically safe, so their working memory remains intact, their creativity preserved. This effort begins with your mouth. Don't weaponize your words. Don't start yelling at your colleagues if you're new at the management game; stop yelling at them if you're a seasoned vet.

5. *Pay close attention to sleep.*

Right before bed, revisit a problem you've been working on. Make judicious use of tea and coffee during the workday if sleep is a problem. Get ready to crash, though. Caffeine is only a temporary fix.

6. *Practice the shift game.*

If you attend a problem-solving meeting, start with whole-group interactions, followed by minutes of disbandment, then reassembly. Populate the groups with people of mixed ages.

7. Create conditions for flow.

The conditions most likely to trigger flow include focusing on the here and now, which means learning about mindfulness. Here, the suggestion is to make mindfulness training available company-wide. We will discuss mindfulness more fully in chapter nine when considering work-life balance.

These suggestions, taken from their evidence-based sources, can change the world. For proof, consider again our Seattle teenagers. The car-counting company they created—called, quaintly, Traf-O-Data—was not a rollicking success. Yet the teens kept plugging along. They eventually changed the focus of the company to software engineering, then changed its name to Microsoft. The teens were, of course, Paul Allen and Bill Gates.

And the rest, as they say, is history.

CREATIVITY

Brain Rule: *Failure should be an option—*
as long as you learn from it.

- If an idea is to be deemed "creative," it needs to be both novel and useful.
- *Divergent thinking* (conjuring up many innovative ideas in an open-ended way) requires a stress-free environment with a long time scale. *Convergent thinking* (conjuring up many solutions to solve one problem) actually thrives in a stressful environment and a short time scale.
- Embrace failure. It is the mechanism that turns something novel into something useful.
- To make failures the most useful to your creativity, try to face them, examine them, and learn from them as soon as you can after committing them.
- Remember that creative failure is not a referendum on you as a person.
- Stop—or don't start—yelling at employees. It stifles their creative output.
- If you are a leader at your company, give your employees a permission structure to fail. Companies that formalize processes that cultivate a failure-friendly environment as well as encourage employees to examine their failures are especially productive creatively.

leadership

Brain Rule:
*Leaders need a whole lot of empathy
and a little willingness to be tough.*

I ADMIT I'M HESITANT to write about leadership. And the biggest reason is that I'm a *chicken*. It takes chutzpah to address a subject with so many uncontrolled variables. I'm not sure that I or my chosen field of study are up to the task. Noted business guru Peter Drucker seems ready to wave a white flag too. Says he: "The only definition of a leader is someone who has followers."

I like his simplicity but not his explanatory power. Surely leadership consists of more than just the ability to attract acolytes. People equally follow bad bosses and good mentors, nasty commanders and nice managers, cruel kings and humble servant-leaders. That doesn't mean these leaders' behaviors are built of the same substance.

This lack of clarity is equally confounding and frustrating, because to reiterate everyone's first lesson in business school, leadership is ridiculously critical to business success. Having a bad boss is the number one reason why people quit their jobs. In pre-pandemic 2018, almost a third of the workforce planned on doing just that. Staff turnover has direct effects on the bottom line too. On average, it takes about $40,000 to replace one employee.

So, what to do? Are there serious attempts to discover the secret sauce of leadership, teachings that instruct bosses how to avoid losing employees at $40K a pop? If the number of books on the subject is any measure, the answer is yes. By 2015, more than 57,000 books with the word *leadership* could be found on Amazon, an astonishing four books published on the subject per *day*. By late 2020, that Amazon number was north of 100,000.

Why so many? Some of these materials declare leadership to be art, not science, which is why there are so many opinions. Leaders aren't born, some say, but rather evolve into their roles via experience. These writers say such learning involves discovering how to balance a constellation of traits, ranging from boldness to empathy, from exclusivity to inclusivity, from not caring what others think to caring only about what others think. How can the behavioral sciences bring anything new to this crowded feast of ideas?

The truth is, they can't—at least not comprehensively. We can, however, bring an important new dish to set at the table, one marinated in evolutionary insight and testable ideas. It's called the prestige-dominance theory of leadership, sometimes referred to as the "Dual Model." It's the subject of the next section, but in no way should the following pages be considered a comprehensive treatment of the idea. I'm not really sure how anyone can compete with 100,000 other books in just one chapter.

I told you I was chicken.

A tale of two generals

We begin our discussion of the Dual Model with World War II, specifically the command reputations of two of its finest allied generals. Their leadership styles could not have been more different from each other.

First up is the legendary General George Patton. Aggressive, intelligent, flamboyant to a fault, Patton commanded his Third Army with an ivory-handled Colt 45 holstered on his right hip, an ivory-handled Smith and Wesson on his left, and two ironclad hemispheres of West-Point-trained brain tissue controlling them both—barely. Nicknamed "Old Blood and Guts," Patton never hesitated to critique subordinates. In a famous speech delivered to his troops, he said:

> *There will be some complaints that we're pushing*
> *our people too hard. I don't give a damn about such*
> *complaints. I believe that an ounce of sweat will save*
> *a gallon of blood.*

Lucky for us, actually. The Nazi-occupied world of the 1940s needed a counterbalancing prizefighter; unleashing Patton into the ring seemed the surest path to a knockout. Some believe World War II didn't go extra rounds because of Patton's aggressive tactics.

His smashmouth personality also got him into trouble, though. He was once sidelined for slapping a shell-shocked soldier who he believed showed weakness. The infantryman was suffering from a condition we now know as PTSD.

Patton was not the only commander with an outsized reputation for leadership. General Omar Bradley, Patton's colleague in the African and European theaters, was also considered a brilliant military leader but for a totally different reason. Bradley didn't have the demeanor of an angry pugilist. He was unpretentious, humble, tactically brilliant, and ferociously dedicated to his troops. He was also given a nickname: "the GI's General." You can see bits and pieces of Bradley's humanity in his leadership writings:

> [N]o commander can become a strategist until he first
> knows his men. Far from being a handicap to command,
> compassion is the measure of it. For unless one values the
> lives of his soldiers and is tormented by their ordeals,
> he is unfit to command.

Years of scholarship have produced a more nuanced view of these men. As you might expect, they did not always live up to their reputations. Yet the stark differences in their leadership styles were real. Having both skill sets present on the same team—if their combustibility could be kept in check—would be invaluable to any war effort. Both personalities are emblematic of the prestige-dominance leadership theory.

Definitions

If we want to make sense of this theory of leadership, we should probably define the word *leadership*. Even given the seemingly wide behavioral disparity of twentieth-century generals, the perspective researchers take on the structure of leadership is as simple as soap.

Sociologists discovered that when people get together, they tend to self-organize in very particular, very measurable ways. The organizing involves the establishment of an asymmetric concentration of power, manifested in this familiar binary: "leaders" and "people who follow leaders." The organizational tendency is so stable, it can be traced to the Stone Age. It's reminiscent of Drucker's simple definition of leadership.

Other leadership models also exist—less hierarchical styles blurring the line between chief and follower, for example—but these have been the historical exception rather than the rule. From ancient pharaohs to European royalty, the binary structure, with the leaders being mostly male, has almost always prevailed. This model provides a convenient, if sometimes depressing, investigative framework to explore not just the definitions of leadership but also ways of studying them—especially if you define leadership in a very particular way.

Lay-audience definitions say that leadership is the ability to persuade others to accomplish something you want done. Science, a bit wordier, comes to the same conclusion:

> ... having a disproportionate influence on collective
> actions and group decisions ... a phenomenon in which
> one individual (the leader) initiates an action and one or
> more individuals (the followers) engage in behaviors that
> match or comply with those initiated by the leader.

The common threads that tie both definitions together are social interactions. That's fortuitous. As we have noted, brain scientists measure social interactions all the time. This means, at least in theory, that brain science could have something to say about how we should lead. But does it? The advocates of prestige-dominance theory sure think so.

Prestige-dominance explained

Prestige-dominance characterizes leadership as a behavioral continuum. The dominance style is populated by leaders who exert authority primarily through brutish strength. They lead through domination and force of will, pushing their agenda onto their followers, usually not caring how the followers feel about it. The opposite end of the continuum, the prestige style, is populated by leaders who exert their authority in a less muscular fashion. They use a combination of wits, judicious communication, and clear regard for the people they lead. The difference comes down to leading with forearms versus foreheads. This idea was published in a paper titled "A Dual Model of Leadership and Hierarchy: Evolutionary Synthesis," the reason it's often called the Dual Model.

Is there a magic mixture of muscles and minds that the Dual Model predicts will work best in business settings? There very well might be. To understand this empirical clarity, we'll need to delve deeper into the two sides of the continuum, both of which are well illustrated in a Christmas movie.

A Christmas Story was as regular a staple in our household during the holidays as baking cookies, and just as sweet. There's a lot to learn about the Dual Model from this confection of a film, two scenes in particular.

The first one concerns the bully of the movie, a kid mischievously named Scut Farkus. Scut was taller than the rest of the kids, physically stronger too, with yellow teeth, a coonskin cap, and a hideous, machine-gun-like laugh. His style of bullying regularly involved ambush tactics. He'd lie in wait by an alley after school, hoping to ensnare gentler souls. Pouncing, he and his toady physically punished his targets for no reason at all. His tortures usually involved putting victims' arms behind their backs, then pushing upward until they cried "uncle." Farkus's leadership was based on bodily strength, with rewards and punishments flowing from physical threats, meted out to allies and enemies alike.

As you might've already guessed, Farkus liked swimming in the dominance end of the leadership pool.

Dominance style

People on this side of the pool obviously derive their power from an asymmetrical distribution of strength. The strength can be physical—like Farkus's ability to overpower weaker boys. The asymmetry can be coalitional too, the ability to compel toadies to do the bidding of the leader. This style is mostly leadership by coercion, exploiting combustible mixtures of anger, fear, and distress to maintain control. Dominance styles often come with loyalty programs too, ensuring control through rewards given to subordinates displaying the most fealty to the dear leader. Those rewards can be powerful, from public displays of approval and respect to more physical manifestations, such as higher salaries and promotions. But it's a cold, binary world that dominance leaders produce. In groups exercising dominance styles, a stark line is drawn between the haves and the have-nots. That line is usually drawn at the discretion of the leader. Most people aren't friends in dominance-run groups: they're allies. And most people aren't opponents: they're enemies.

Dominance leaders can make life miserable for subordinates and, without tempering themselves with other behaviors, often do. But dominance styles exist for a reason. These leaders can marshal resources quickly, which is especially valuable when decisions require reflexive, don't-argue-with-me responses. This style is useful in emergencies, such as fighting off enemies, dealing with freeloaders, or handling intragroup conflicts, especially if those conflicts pose direct threats to central power.

Joseph Stalin was a leader whose use of dominating tactics showed the extreme side of this leadership style. His desire to cling to power mobilized the industrial might of the Soviet Union, keeping the marauding armies of Nazi Germany at bay. Yet that same desire led to the deaths of millions of innocent people both before and after the war.

Research shows that the punishing behaviors of dominance styles are not good at sustaining long-term productivity, especially if dominance-style leaders are tasked with maintaining innovative leadership. (You could insert the entire chapter on creativity right here.) To understand what alternatives exist, we have to swim over to the other side of the Dual Model pool, under the sign marked "prestige." To explain this style, we turn to our second example from *A Christmas Story*.

Prestige style

The movie's dinnertime scene involves the main character's little brother, Randy, a lad not more than five or six years old. He is refusing to eat his meat loaf and mashed potatoes and gravy.

"Alright," his father growls, channeling his inner dominance. "I'll get that kid to eat. Where's my screwdriver and my plumber's helper? I'll open up his mouth and *shove* it in!"

Mom quickly intervenes and gently asks Randy, "How do the little piggies go?" Randy suddenly lights up and snorts. He starts to laugh. "That's right!" Mom says, sensing a breakthrough. "Oink, oink! Now show me how the piggies eat!" Mom points at the uneaten food on Randy's plate. "This is your trough. Be a good boy and show Mommy how the piggies eat!"

Randy immediately attacks his food, diving face-first into his plate, no hands, squealing like a pig. Mom breaks into a gale-force laugh as Randy smears potatoes and gravy all over his face. "Mommy's little piggie!" Mom laughs as Randy finishes his dinner. Mission accomplished.

This delightful scene is a terrific example of how dominance leadership contrasts directly with prestige leadership. Whereas people who use dominance styles rely on asymmetric distribution of muscle to accomplish their goals, people who use prestige styles rely on the asymmetric distribution of insight. Mom had *knowledge* about what it would take to get her little boy to eat his food, which she implemented—and it didn't involve a screwdriver. Some would call this wisdom.

Prestige leaders possess the skills and knowledge necessary to understand the relational ecologies of the people they lead. To motivate their followers, prestige leaders identify what makes them tick, then use this insight to accomplish their goals. This subtlety is often missing from dominance-only styles. Prestige leaders seem to intuit that fear, anger, and brute force are best used as tools of last resort. If you suspect prestige leaders are marinated in pro-social Theory of Mind, you are right on the money.

Prestige leaders are imbued with other behavioral characteristics besides high RME scores. They willingly share their resources with the followers in their care. They seem to be interested in their followers' fulfillment as well as their own. Prestige leaders' incentives are positive, as opposed to the incentives of the dominance model, which are largely negative. Dominance leaders tend to command; prestige leaders prefer to influence.

The reason it's called prestige

Leaders who practice this stable marriage of wisdom and generosity usually garner powerful positive reputations. They accumulate prestige, attracting people like moths to light. Indeed, people who practice prestige leadership don't usually have to recruit people to follow them. People follow them of their own accord.

What their subordinates are mostly attracted to is the familiar promise of relational safety. It is, after all, a potent thing to know that you are understood and that your hard work will be rewarded. Prestige leaders generate such forceful interpersonal gravity that their relationships with followers often become personal. People *like* them, want to orbit around them, and may want to emulate them. In many instances, these leaders are called *charismatic*.

That's not to say people aren't sometimes attracted to dominance-style leaders too. A leader's ability to make clear, muscular decisions can be refreshing to followers when life becomes complex, ambiguous, or threatening. When the leader's personal power exacts real results, people can be grateful for the strength. It might be pleasant to be around insightful, empathetic people, but those are the last traits you need in the heat of battle.

So, here you have two sets of leadership styles: one dependent on fists, the other dependent on fame. Which does science say is the best one?

What science says is that this is the wrong question. The most effective leaders have both abilities in their executive toolkit. They can figure out when to take one out and when to put one away.

But research also says that one style is needed far more than the other. Studies show that business-related conflicts—ones that might actually require a leader's inner Patton—just don't happen very often. That means the need for dominance styles are relatively rare. More common—and more important—are the mundane, day-to-day decisions executives and managers face. These decisions daily require dozens of incremental wisdoms, whose cumulative effects drive companies forward. Such prestige-ready wisdoms do not require a heavy hand. They require a dexterous hand. "Show mommy how the piggies eat" is almost always a better first strategy than getting someone to cry "uncle."

Dualing models

The Dual Model continuum isn't the only management idea out there, of course. Some researchers organize leadership into the idea of *personalized power*, for example, which is authority flowing from a person's own abilities. This contrasts with *positional power*, which is influence flowing from simply being in a position of authority.

Most of these other models find themselves routinely referencing, incorporating, and balancing the same prestige/dominance behavioral elements as the Dual Model. As a scientist, such consistency gets my spidey-senses a-tingling.

Consider research done by James Zenger, whose team asked 60,000 employees: "What turns a boss into a great leader?"

Zenger examined two characteristics particularly closely: results focus and social skills. Bosses who were focused on results achieved the goals they set out to achieve and met the deadlines they needed to meet, generating the quality of goods and services they promised to produce. Bosses who practiced good social skills could communicate to their employees with clarity and empathy.

These characteristics had to work in tandem for bosses to make the grade. If an executive or manager possessed results-focused instincts but the social skills of a spreadsheet, only 14% of those surveyed felt that turned the hack into a hero. If the boss had the social skills of a kindly saint but couldn't keep a team focused and productive, only 12% felt that was sufficient for canonization. The situation completely changed when both characteristics were resident in one person, however. According to Zenger, if managers were results-oriented *and* possessed the disposition of Mother Teresa, a whopping 72% of respondents felt they were in the hands of a great leader.

This combination of (1) drive to achieve a goal and (2) presence of mind to get employees to buy into said goal sounds suspiciously similar to the prestige-dominance template.

Business author Greg McKeown bumped into the same thing whilst looking in exactly the opposite direction. He wanted to know what behaviors employees associated with bad managers.

After interviewing 1,000 employees from top American companies, McKeown discovered a binary: About half the people said that the worst bosses are too controlling, too dictatorial, too willing to micromanage their employees' everyday work lives. These bosses were christened "over-managers."

The other half of the people said the opposite: that the worst bosses weren't involved enough, possessed poor accountability skills, and gave virtually no feedback. Most were "nice enough," but their agreeableness revealed more of a desire to avoid conflict than to manage effectively. These bosses were christened "under-managers."

The best ones? The ones that took the middle way, which simply means they had both behaviors in their toolkits and knew when to use them. You can easily run into these twin ideas no matter where you go.

I can still feel my spidey-senses tingling.

All about babies

Given the Dual Model's consistent ability to make guest appearances in other people's management theories, does this model point to something innate to us as humans? Is our habit of self-organizing into a prestige-dominance continuum carved into our biology?

The answer is "nobody really knows." Yet we have tantalizing hints that certain social tendencies aren't entirely due to certain social forces. For example, you can detect very early in life—by *early*, I mean babies—consistent responses to social power. Infants appear endowed with behavioral templates concerning how humans are supposed to interact, templates detectable in children as young as twenty-one months old. These templates appear to be made of prestige and dominance expectations. Said one researcher:

> *... young children already have cognitive templates*
> *for each leader type ... followers and outside observers*
> *(even when they are infants) easily distinguish between*
> *prestige- versus dominance-style leaders and apply*
> *context-sensitive preferences for each type.*

How do we know? Understanding the mechanics of how researchers measure infant cognition might be instructive here. It's been long established that infants—like adults—gaze longer at an input if they detect some difference. Suppose a baby sleeps in a room with two windows, behind each of which is a tree. If the tree is removed from one window but not the other, the baby will stare longer—and harder—at the now-treeless window. Gazing is a reliable way to test what infants are paying attention to.

In one famous social-power experiment, twenty-one-month-old infants watch a scripted play, with adult leaders interacting with followers in specific ways. One leader exhibited prestige-style behaviors while working with subordinates, just like the mom in *A Christmas Story*. The other exhibited more dominant, authoritarian-style behaviors, like good old Scut Farkus. Astonishingly, babies could tell the difference. They expected subordinates to react in specific ways, depending upon the perceived leadership style.

Researchers next allowed the babies to observe a specific interaction, a leader giving an order to a follower, with the leader then exiting the room. Using a sophisticated version of the gazing assay, the investigators were able to ascertain what the baby expected the follower to do next. If the leader exhibited prestige behavior, the infants expected the follower to carry out the order even in the leader's absence. But if the leader exhibited dominance behavior, the infants expected the follower to carry out the order only when the leader was physically present. If the leader was absent, disobedience was actually expected. Babies seem to have resident in their brains the perception that "when the cat's away, the mice will play."

The researchers concluded that the babies were referencing an embedded social template, complete with predictions about what people were likely to do in response to specific types of leadership behavior.

This is just one experiment in a dog-pile of data showing that babies possess cognitive models concerning how humans interact with each other. And they tend to follow Dual Model house rules. Yep, management theory, pediatric edition.

As adults

Exactly how do those templates emerge? As mentioned, nobody really knows (feel free to cue up the nature/nurture issues we discussed in the introduction to this book). It's possible we were born with them. It's also possible we were not born with them. We only know they're detectable at ages when toddlers get their molars.

These templates have also been studied in children as they leave infancy and enter their school years. Researchers *still* find evidence of the Dual Model—and its social impact—in school-age children. It's true that the most well-known students in class are often the leaders. But the most *popular* kids in class are those who exhibit both styles of the Dual Model and are fully capable of shifting between them.

Sadly, this research also shows that children raised in punitive authoritarian environments don't freely shift between the styles at all. They become socially aggressive, displaying something called "externalizing behaviors" (a euphemism for the word *bullying*). They grow up to prefer authoritarian leaders as adults. If they themselves become leaders, their preferred leading style is dominance.

Which leadership template executives ultimately choose has obvious consequences for their surrounding relationships. Their behavior affects both home and work, but precisely what are those consequences? Exactly how do these styles impact the business setting? What can we do next Monday to keep the behavioral ledger balanced away from bullying and toward productivity?

We're going to address these questions using a teaching device common to both medical and business schools: a case history. Our subject will be the short, famous history of the Enron Corporation of Houston, Texas. Its breathtaking rise and gut-wrenching fall is attributed to the contrasting leadership styles of two men who filled the CEO chair: Richard Kinder and Jeffrey Skilling. This is an ugly tale but extremely instructive. It underscores both the strengths and weaknesses of prestige and dominance styles as well as the consequences of those strengths and weaknesses.

Some quick background: Enron was an energy company, the 1985 offspring of the marriage between InterNorth and the Houston Natural Gas Corporation. Richard Kinder was brought in to nanny the young corporation as president and COO during its formative years while the merger's father, a man named Ken Lay, left to go schmoozing in Washington, DC.

Enron up

Kinder was well named. He displayed a remarkable talent for prioritizing the welfare of his employees, even making their personal details his personal business. By all reports, he wasn't relationally controlling, just caring, promoting what one observer called a "family-like atmosphere" at work. He clearly had been schooled in the saintly Mother Teresa component Zenger described. His behavior reflected important elements of the prestige style of leadership.

But that wasn't the only managerial gadget in Kinder's toolkit. Realizing there are differences between workplace colleagues and family members (you can fire a colleague, for example, but you can't fire your brother), his managerial style included harder edges. He enforced a company-wide work ethic aimed at achieving goals, making deadlines, and producing quality services. He used his prodigious memory to constantly surveil how each unit in his growing enterprise was performing. Insisting on transparency, he was quite willing to challenge his executives if he saw something wrong, then ask his managers to do the same with their subordinates. This habit earned him the name "Doctor Discipline."

Yet even in the midst of this results-focused behavior, his relational skills glowed. Classic prestige-esque consequences flourished. Kinder's intolerance for secrecy became a virtue, imbuing the company with that rarest combination of traits: voluntary trust and honest accountability.

Enron's bottom line was also an indication that Kinder's leadership style was working. Enron's most profitable years occurred under his watch, with earnings skyrocketing to $584 million from $200 million, on the back of revenues soaring to $13.4 billion from $5.3 billion.

Which made it all the more puzzling when Kinder was replaced. The board forced Kinder out in 1996. Many articles, books, and documentaries have chronicled this transition, not only for its suddenness but also for its tragic consequences. Enron hired a replacement who sailed this thriving corporate titan right into an iceberg. The ship sank in 2001.

The villainous skipper was Jeffrey Skilling, a leader right out of dominance-style central casting. According to one source, his actual influences were Darwinian. He believed that corporations were best run as dog-eat-dog, survival-of-the-fittest enterprises.

Consistent with this perspective, Skilling immediately put into place an employee-review system called the Peer Review Committee. This system ranked all employees on a 1-to-5 scale, 1 being the best, 5 the worst. *Regardless of competence.* This evaluative main dish also came with a side of public humiliation. All performance reviews were posted on the company's website, complete with a picture of the assessed person. It also came with an occupational death sentence. An employee with an evaluation of 5 could be fired outright or given two weeks to find another position within the company. Even if they were really good at what they did, if their fortnight search proved unsuccessful, they were sent packing. Under Skilling's helm, meritocracy was for sissies.

Enron down

As you might imagine, the nervousness employees felt about their job security began to tear a hole through Enron. Employees started seeing coworkers not as colleagues but as competitors. Backstabbing became the norm. Managers weaponized the evaluation system, dangling positive ratings as rewards for personal loyalty rather than as honest assessments of productivity.

The aggression even contaminated attitudes toward Enron's customers. One executive was clandestinely taped celebrating a California forest fire that had shut down the local power grid, giving Enron an opportunity to jack up energy prices: "Burn, baby, burn. That's a beautiful thing," the executive declared.

Coupled with Skilling's penchant for taking high risks, Enron's corporate hull began taking on water, mostly in the form of debt. Executives tried to hide the flooding at first, attempting to deceive regulators, but the corruption was eventually discovered. Enron declared bankruptcy in 2001, and several executives, including Skilling, served time in jail. It was quite a fall. At its peak, Enron's shares were pegged at $90.75. At bankruptcy, they went for 26 cents.

The bottom line? Leadership styles have consequences. Enron didn't fire the entire organization. They just replaced the head, but that made all the difference in the world.

What to do next Monday

Though your company might not be hobbled with Enron's extreme leadership problems, maybe you see uncomfortably similar debilitating tendencies in leaders with whom you are currently associated. Or perhaps you're an executive or manager who has embraced more dominance-style leadership behaviors than you'd care to admit. We are going to spend the last few sections of this chapter discussing how to avoid some of the mistakes that leaders like Skilling commonly make. We, thus, begin our journey of planning what to do next Monday a bit early. This journey will involve a fair amount of writing.

I start with a basic mistake that Skilling, the Enron board, and its executives made about Darwinian behavior. Skilling only ingested part of the Darwinian story: the socially selfish part. Missing entirely was the socially selfless part, which takes pleasure in others succeeding. Scientists still debate whether the major driver of evolution is competition, cooperation, or some uneasy mixture of the two. Executives who embrace only the selfish part of their Darwinian roots can have crippling impacts on others, as Californians can attest.

Enron's disastrous value system based on self-centeredness shows us, oddly enough, what to do next Monday: figure out how to become sustainably less self-centered. Brain science says a great deal about how we should go about doing that. We will look at two studies. The first study looks at what keeps people from becoming clinically depressed. The second one attempts to understand what makes people authentically happy. Both lead to a completely unexpected place: the formal study of human gratitude.

An attitude of gratitude

Researchers usually characterize gratitude as the coupling of two realizations—one easy, one hard—whose union gives birth to a positive emotion. The easy realization is the recognition that something good has occurred. The hard realization is discovering that the goodness came from a source outside yourself. If you can achieve the second realization, an emotion consisting of satisfaction showers over you, one that scientists often describe as *warm*. You realize your soccer team just won the game, but it was your teammate rather than you who kicked the winning goal. If you can nevertheless be happy for the outcome, appreciating the efforts of your teammate, warm feelings ensue.

Put a circle around that second realization, the source-outside-yourself one. That's the propulsive beginning of how people learn to be less self-centered. Put another circle around the ensuing positive emotion. That's the reward your brain dispenses for convincing you that you're not the center of the world. Research demonstrates that if you can persuade someone to refrain from being self-referential and convince them to be grateful instead, positive relational dividends begin accruing—including the potential to be a very productive leader.

The first traces of this interesting association were discovered in studies of psychiatric disorders. Gratitude development was found to be a force-multiplier for patients receiving therapy for clinical depression. Gratitude helped untie the knots in their brains, reducing both the duration of their depressive episodes and the frequency at which they occurred.

Gratitude was also found to have extraordinary benefits for people not in therapy, benefits that directly affect leadership ability. It enhances people's ability to feel empathy, for example, while tamping down feelings of envy, resentment, and aggression. It also gives people the remarkable ability to decrease their retaliatory desires when responding to perceived enemies.

People who cultivate gratitude also acquire an advanced ability to make friends and keep them. It's redolent of the old adage that says that if you want to make friends, first be one. Grateful people are more outwardly focused on others—there's that selflessness again—and more thoughtful in their everyday social interactions. Continuous exposure to grateful people incites in others a desire to stay socially affiliated, strengthening high-quality, long-term social bonds.

Gratitude also affects how people process stress. Being consistently grateful not only relieves the tension in your life but also makes you more stress-resistant when bad stuff happens. The findings are robust enough to be observed in soldiers afflicted with PTSD: people with highly cultivated, habitual feelings of gratitude recover more quickly from trauma they sustained on the battlefield than those without this perspective.

The neurobiology of gratitude

I've run smack-dab into the power of gratitude in a Starbucks drive-through, and I'll bet you have too. It's that pay-it-forward experience, in which somebody in front of you pays for your drink as well as their own, and you don't find out about it until it's your turn to pay. The feelings of warmth and gratitude always wash over me, and I invariably reciprocate the kindness for the next person.

These feelings have a solid neuroscientific understory. At least three networks of neurons are involved in the brain when I—and you—experience gratitude.

The first network traffics in a neurotransmitter you may have heard of before: serotonin. It involves a brain region you may never have heard of before: the anterior cingulate cortex (mercifully shortened to ACC). Serotonin is a Swiss Army knife of a biochemical, but its largest job is to promote contentment and mood stability. (Indeed, people who suffer from depression often have problems with serotonin regulation.) The ACC releases serotonin when you feel grateful. Interestingly, it is also involved in decision-making, assisting you in evaluating—even forecasting—the outcomes of a given action.

The second region involves another neurotransmitter you're probably familiar with: dopamine. Dopamine is the brain's archetypal feel-good chemical, directly involved in feelings of pleasure and reward. When gratefulness comes over you, regions near your brainstem give off a small but very powerful squirt of this pleasant chemical. It's one of the reasons you feel pleasure when you're thankful.

So, gratitude is coupled with the two most powerful feel-good neurotransmitters in the brain's arsenal. You become more content, then get rewarded for being more content. There's a delightful boomerang effect here: by not focusing on yourself to obtain your own happiness but instead focusing on other people's happiness, you get your own happiness back anyway. There's little wonder why gratitude effects the human experience so deeply.

One last region involved in mediating gratitude may be the most interesting of all, certainly the most tongue twisting: the intraparietal sulcus (a region just above your ears) and the inferior frontal gyrus (a region just in front of your ears). Their portfolio? Mental arithmetic. These regions are involved in quantifying your daily world, calculating things like *how much* and *how little*. No one really knows why they get recruited when you experience gratitude. It's possible you use them as a balance sheet, keeping mental tabs on what you owe to people and perhaps, ominously, what people owe to you. It also may be due in part to the way gratitude is measured. Many gratitude experiments involve a subject unexpectedly getting a reward from someone, often monetary in nature, while the subject's brain is being imaged.

Whatever the role, gratitude research shows the amazing power of getting over yourself. From forming positive relationships to being stress resistant, the benefits read almost like a how-to manual about cultivating prestige-style leadership.

Start writing

So, how do you create consistent gratitude? It's one thing to be grateful once, but the literature shows something rather uncomfortable. Though surprisingly stable benefits are still measurable with limited practice of gratitude (as we'll see in a moment), the habit of gratitude needs to sign a long-term lease in your brain to exert long-lasting effects. Developing a durable, reflexive "attitude of gratitude" is key.

Practical exercises to this end have been tested in real-world settings. They really do work, but you have to practice them for a long time before their effects become measurable. A surprising number involve writing something down.

The first writing exercise involves creating a gratitude journal. You visit the journal daily, like you would a diary. The entries can be a list of people, events, or circumstances that you experienced that day for which you are truly grateful. You can write down as few as one entry or as many entries as you want. (Researcher Martin Seligman suggests three.) Your writing is rendered more potent if you also explain why the person, event, or circumstance meant something to you. If a competitor gives you a sincerely friendly hand-shake for example, you might write, "That meant a lot, because I thought we were enemies."

The second exercise is to get in the habit of reflexively writing thank-you notes. This can be done with dead trees and ink or with cell phones and texts. It can also be done in your head. When someone does something nice for you, mentally thank them, even if you can't type it or write it down.

The third habit is a long-form extension of the thank-you note. Get in the habit of writing gratitude letters that you follow up with gratitude visits. Think of a person who has meant something to you and write them a letter (three hundred words is a good length) describing their impact on your life. Then visit them, if possible, and read the letter aloud. If executed correctly, you should also pack a handkerchief.

Note that, in each case, the advice is to do something tangible, physical—something you can hold in your hands, read with your eyes, do with your voice. There is a reason for this approach. People feel almost universal opposition whenever they try to evict themselves from the center of their universes. Being successful at it takes lots of new learning. One of the most powerful ways to cement that learning is to couple a cognitive assent with a motor skill, a multisensory idea we'll discuss later. The goal here is to get started immediately and still be doing it three months from now.

One important consequence of writing is that it creates a tangible paper trail. You can look back on your efforts and chart your progress whenever you feel rebellious, discouraged, or weary of trying to be better. Writing may, thus, function like verbal epoxy, helping to solidify the experience.

Business implications

Studies about gratitude's ability to force-quit our self-centered tendencies have been going on for decades. These efforts include subjects of direct interest to business professionals.

Leaders who regularly express gratitude to their subordinates cause those subordinates to be more productive, for example. One remarkable study examined workers involved in alumni fund-raising. Employees made 50% more calls to potential donors if their managers regularly expressed gratitude toward them, as opposed to the controls, whose managers were instructed to express no gratitude.

Such changes in productivity are common, and we may know the reason why. Leaders who regularly convey gratitude cues engender feelings of being valued in subordinates. These feelings snowball into employees feeling inspired to do their best work, which results in higher job satisfaction. Such satisfaction is the magic elixir that keeps employee turnover low.

Gratitude also positively affects the people who regularly express it, especially on their mental health. Surprisingly, a little goes a long way. Even when tangible activities (such as the writing exercises) were halted, mental health benefits were still measurable twelve weeks later.

So, how do these exercises affect leadership? The effects of gratitude are like an icebreaker, crashing through your crusty selfishness and making a path others can follow. And that's great for business, for nothing eases the way toward productivity more than being able to follow a clear, welcoming, obstruction-free path. In my view, people end up following you not because they have to but because they want to, one of the hallmarks of prestige-style leadership.

What to do next Monday is, thus, easy to understand. Learn from Enron and the generals of World War II: find the right mix of prestige and dominance in your leadership style, limiting dominance to an as-needed basis. Perhaps you discover your leadership is freighted with an overabundance of dominating behavior. The best way to rebalance is to confront the self-centeredness directly. You can confront it by practicing gratitude, using any of the techniques listed here. With practice, you can move away from being the overbearing center of your relational universe. You could start with some great scientists—dozens strong—who discovered how easy this is to do.

LEADERSHIP

Brain Rule: *Leaders need a whole lot of empathy
and a little willingness to be tough.*

- Leaders practice leadership on a sliding scale according to the prestige-dominance model, employing a mix of strength and force (dominance) with insight and empathy (prestige).
- The most effective leaders possess both dominance and prestige capabilities and know when to deploy them—prestige for the majority of everyday tasks, and dominance for sporadic conflicts, emergencies, or situations that require efficient and concise decision-making.
- Keep dominance-style leadership behaviors to a minimum. Too many can create fearful and often miserable work environments for employees.
- Make a concerted effort to be grateful, especially toward others. When gratitude is present in people who hold leadership positions, subordinates are more productive.
- Make an organized effort to be grateful by consistently writing down what you are thankful for, over a sustained period of time. It can enhance your ability to feel empathy.

power

Brain Rule:
*Power is like fire. It can cook your food
or burn your house down.*

POWER DOES FUNNY THINGS to people, though it might be just as accurate to say power also *reveals* funny things about people.

Take the 1970s Ugandan dictator Idi Amin. His reign was notorious for its corruption, eccentricity, and brutality. He more than earned his nickname: the Butcher of Uganda. Amin was murderous (he killed hundreds of thousands), extremely sexually active (he went through six wives and may have fathered up to fifty-four children), and possessed such ribald behavioral quirks that *Saturday Night Live* satirized him no less than four times. He created the following official title, to be read whenever his presence was announced:

> *His Excellency, President for Life, Field Marshal Al Hadji Doctor Idi Amin Dada, VC, DSO, MC, CBE, Lord of All the Beasts of the Earth and Fishes of the Seas and Conqueror of the British Empire in Africa in General and Uganda in Particular.*

He also claimed he was King of Scotland.

The effects of power on people have been depicted in fictional narratives from *King Lear* to *Citizen Kane* to more recent lighthearted fare like the TV show *The Office*. One episode of the show, entitled "The Coup," features the character Dwight Schrute—a power-lusting wannabe executive played by Rainn Wilson—christening himself Assistant Regional Manager as he guns for the position of his actual Regional Manager, Michael. Dwight plots with his office-crush, Angela, to take control, but Michael discovers this takeover scheme and decides to employ a ruse. He tells Dwight he's stepping down, handing him the levers of power. Dwight soon makes a wreck of things, planning the firing of employees left and right and scheming with a colleague about future managerial conquests. People, naturally, are shocked. Michael reveals the ruse, relieving Dwight of his responsibilities—and his plans. It's obvious these tendencies were always resident in Dwight, but when he obtained power, they came out at full throttle, revealing who he was—maybe even, given enough time, *changing* who he was.

Though Amin and Schrute come from completely different narratives, only the last story is fictional, and just barely. What Amin and Schrute have in common is a preoccupation with themselves and their power, a tendency to ignore the welfare of people they're leading, and a power-suffused sexual appetite. Their stories are emblematic of the vulnerabilities to which virtually anyone can fall prey when they acquire power.

How do we explain this bad behavior? Why do some leaders rise to the occasion while others declare themselves Lord of All the Beasts? What is it about power that can turn people into monsters—or perhaps, more accurately, reveal the monsters resident in them all along? And what can we do to prevent power from going to our heads if we acquire it?

We can address those issues after we define a few terms, beginning with the word *power*. The definition we use comes from psychologist Dacher Keltner, a researcher who has studied the effects of power for years:

> *In psychological science, power is defined as one's*
> *capacity to alter another person's condition or state of*
> *mind by providing or withholding resources—such as*
> *food, money, knowledge, and affection—or administering*
> *punishments, such as physical harm, job termination,*
> *or social ostracism.*

Notice that this definition describes the ability to control not only physical goodies but also cognitive ones. Some leaders want control over what people have, others over what people think. That's why this definition, addressing body and soul, is important.

But we won't just use subjective definitions in our exploration of the neuroscience of control. Researchers have also devised quantitative ways to study it. One common method business schools use has to do with a concept called *social power*. Social power can be described by a formula: combining a person's net worth, yearly income, education level, and occupational prestige into a single number. This approach takes into account that for most people, money—no surprise here—is related to power. The higher the number, the more social power one has.

Social domination, mental influence, high net worth—regardless of what you study, power is all about controlling resources. We'll use Keltner's definitions and the idea of social power to identify what real control does to real people. Spoiler alert: as the "King of Scotland" illustrates, it's not always a pretty story. But as the Assistant Regional Manager illustrates, it is sometimes a funny one.

Of jails and joules

What is it about the ability to control thoughts and things that's so desirable to so many people? What behavioral changes are revealed when people acquire power? The behavioral research world knows, mostly from the results of two of the most controversial experiments in all of psychology. The efforts were bicoastal—one experiment done at Stanford, the other at Yale.

The California experiment was run by legendary psychologist Phil Zimbardo in 1971 and was called the Stanford Prison Experiment. Zimbardo investigated what happened when typically functioning undergraduates played in a make-believe jail. Some of them were assigned to be guards and given all the relational power, while others were assigned to be prisoners, recipients of whatever behavior the guards deemed fit.

Zimbardo didn't know it, but his experiment was about to plunge off a cliff. After forty-eight hours, the "guards" began abusing the "prisoners," first mentally, then physically. The abuse got so bad, the experiment was halted after only six days (it was supposed to last two weeks). Power, even in make-believe situations, had the ability to quickly corrupt people.

As you might expect, this experiment proved to be very controversial. Objectors noted that certain aspects of the original work couldn't be replicated and pointed to ethical violations of the data that could be repeated. Despite the disputes, the disturbing core idea remained. Authority changes people, and the changes aren't always for the better.

Whereas Zimbardo's study focused on the behavior of the *powerful*, Yale researcher Stanley Milgram's experiments in 1963 studied the behavior of the *powerless*. Milgram conducted his experiment twenty-six times, using research subjects of varying ages. Each iteration involved three people: an authority figure (in this instance, a scientist), a lab "confederate" (a paid actor), and the research subject.

The experiments started out with a lie: the subjects were told they had signed up for a memory test between them and the (unbeknownst to them) confederate. Every time the confederate got a wrong answer, the subjects were to apply an electric shock by pressing buttons of increasingly powerful voltage. The final button was a death-dealing 450 volts, indicated with a skull and crossbones. The subjects couldn't see the actor but could hear him and his reactions to the shock.

At first, the actor would indicate mild discomfort when the subject pressed the button (an "ow"). But as the shocks increased in voltage, so did the verbal agitation. The actor eventually begged for the experiment to stop, even screaming at the end. At the 450-volt level, no sound was heard. There never were shocks, of course.

Milgram was just interested in how many people would press the skull and crossbones button.

The answer was depressing. Almost 65% did.

These experiments proved to be extremely provocative, and not just because of the conclusions. Attacks were made on their methodologies, interpretation, ability to be replicated, and especially their ethicality. But the effects were clear: Power was doing something weird to the human brain—or perhaps, more disturbingly, revealing something about the human brain. None of it was funny.

The problem with net worth

These experiments admittedly describe extreme examples. What about less intemperate, more everyday illustrations, like when somebody gets promoted, acquires access to resources, then becomes capable of granting that access to others? Are measurable changes still detectable?

Turns out the answer is yes, complete with this core finding: power creepily starts causing people to favor self-interest over the interests of a given group, a process researchers call *disinhibition*. Power actually does to leaders the opposite of what leaders are supposed to do—even to the point of compromising ethics.

Need proof? A paper published by the National Academy of Sciences will do nicely. It examined behaviors of high-net-worth individuals, used to wielding the levers of power. These behaviors were compared with the everyday behaviors of low-net-worth individuals, who weren't used to wielding anything. All were observed both in the lab and in "the wild."

It makes for distressing reading. In the lab, high-net-worth individuals were more likely to cheat at games of chance than low-net-worth individuals were. They were more willing to lie under experimental conditions when given a choice—also compared with low-net-worth controls. Under laboratory conditions designed to measure greed, they were demonstrably greedier: subjects with high net worth took more candy from a candy jar than their poorer colleagues, even when informed that any leftovers would be donated to kids in a nearby building.

The list goes on. The people with high net worth were more likely to lie during negotiations, cheat when they believed they could win a prize, and take goods for themselves even when these were of value to others.

And just to dog pile on the experimental work, observational studies of high-net-worthers in "the wild" exhibited similar self-oriented habits. Rich folks were more likely to break the law while driving, for example. Specifically, they were more likely than poorer folks to cut off drivers at intersections and cut off pedestrians at crosswalks (30% versus 7%).

This isn't a fun group, but they do tell an interesting story, one increasingly fleshed out in a behavior remarkably close to mind reading. You know the behavior already. It's the aforementioned Theory of Mind.

Loss of Theory of Mind

There are many scientific ways to show the relationship between being powerful and being obnoxious. It has to do with losing Theory of Mind skills. Researchers like Adam Galinsky have demonstrated, for example, that even casual brushes with power create measurable fractures in emotion-detection abilities.

In one investigation, Galinsky asked experiment subjects to recall a time when they felt power over another (called the *experiential prime*). Control subjects were asked to recall what they did yesterday (the *neutral prime*). Both then took sensitive psychometric tests measuring how well people detect emotions, similar to the RME described previously.

Galinsky found that, on average, the power-primed group produced 46% more errors than the controls. They also became less sensitive. Amidst a forest of other experiments, he concluded that power is linked with a decreasing ability to detect emotion accurately, and suggested that power is associated with diminished perspective-taking.

Two things to note here: First, the sensitivity loss was ridiculously inducible. The simple *memory* of having power over people was enough to change behavior. That sensitivity is a phenomenon we'll run into again and again, by the way. Second was the use of the words *diminished perspective-taking*.

Those two words represent the signature behavior of Theory of Mind. That point is important, because it suggests something testable by neuroscientists who image neural tissue for a living. From their efforts, we know where Theory of Mind behavior arises in the brain. It is mediated by a series of neural circuits called the mentalizing network. This network is composed of a broad range of hard-to-pronounce neural areas, from the dorsal medial prefrontal cortex (a region behind your eyes) to the awkwardly named precuneus (a region near the top of your head).

Which leads us to a very important question: As people gain power, can we drop in on their brains to see how their mentalizing networks are functioning? Is the reason why increased social power reduces the ability to take others' perspectives because the mentalizing network is, metaphorically, short-circuiting?

The answer is yes. The title of the paper from neuroscientist Keely Muscatell, then at UCLA, says it all: "Social Status Modulates Neural Activity in the Mentalizing Network." The conclusion is breathtaking: power literally rewires the brain.

The power of empathy

So, power affects Theory of Mind. Power also affects a cognitive gadget that sounds a lot like Theory of Mind—complete with its own "layperson" definition. That gadget is empathy. To understand what power does to empathy, we need to define empathy. We'll use the layperson definition for the sake of this section: "The ability to recognize and share the emotional space of another person," as if you were experiencing what another person was experiencing yourself. It's embodied in the old idiom to not judge people until you've walked a mile in their shoes.

Consider this example of a mother strolling through a mall with her toddler and two-week-old infant. She tries to navigate getting food while pushing a stroller. Her toddler is tugging at her shirt while her infant is crying, and she realizes that she has forgotten the extra diapers for her infant. She suddenly feels overwhelmed, sits down, and puts her head in her hands. A nearby older lady, who has never met the mother, sees what is happening. She can read the distress of the mother, hear the complaints of the toddler, and deduce the situation. But something else is also happening to the older woman. She *feels* the stress of the mother, almost as if she is taking the stress on as her own. The older woman gets up and rocks the infant in the stroller while the mother gets her toddler's food ready. The older lady says, "I know it's hard now, but it'll get easier." She hands the stroller back to the mother and leaves as suddenly as she appeared.

This generous stranger was obviously exercising empathy. The research community recognizes two varieties of it. The first variety is *cognitive empathy*, the willingness to understand the emotional experiences of another (some researchers think this is good old-fashioned Theory of Mind). The older woman displayed her capacity for cognitive empathy by reading the mother's face, body position, and change in voice. The second variety of empathy is *affective empathy*. This was the type of empathy the woman could feel, almost as if she were walking in the mother's shoes.

Researchers discovered that powerful people have reduced empathic ability, regardless of the variety measured. They brought out the big, biological guns—imaging neuroscience—to explain.

In one experiment, researchers looked at empathy centers in the brains of people of varying net worth. These brains were examined while their owners gazed at heartbreaking pictures of kids in a cancer ward. The poorer the subjects were, the *more* activity their empathy centers displayed. The wealthier they were, the *less* activity those same empathy centers displayed.

The previously mentioned neuroscientist Muscatell found that people with high net worth couldn't easily understand the interior motivations and intentions of others. This lack of empathy could be detected in the children of socially powerful people, beginning around age four.

Researchers at Arizona State detected changes in empathy observable simply by looking at changes in scalp surface electricity (brain waves). Through ERP (event-related potential) technology—a way of measuring surface electricity, using something that looks like a hairnet—these researchers found that empathy networks were simply less active in powerful people. They also showed that powerful people were almost completely unaware of their deficit, feeling they were as empathetic as anyone. Neuroscience clearly showed they weren't.

If rich, powerful people were *less* empathetic, were less powerful people *more* empathetic? The answer is yes. Researchers showed that less socially powerful people were more accurate on empathic-accuracy tests (yes, that's a thing) than were powerful people. Those with less social power were also more accurate at evaluating emotions of others during real-world interactions.

Researchers have gone further than just looking at changes in surface electricity on the scalp to reveal these data. They've also done internal, real-time imaging of the brain. The results are sadly consistent, and thoroughly instructive.

Mirroring empathy

Can the brain be imaged in the act of being empathetic? It's astonishing to say this, but the answer is yes. Researchers deployed noninvasive imaging devices such as fMRIs (devices which use external magnets to measure blood flow changes in the brain) as a way to detect interior activity. They trained this powerful technology on what's going on in the brains of people who behave empathically, and found a suite of neural networks, which they named after something one finds in most dressing rooms: mirrors. Because of their reflective power, these remarkable circuits are called *mirror neurons*.

Mirror neurons are specialized neural networks. They react to information from the outside world in a very peculiar fashion. They actually mirror the activity of that action, as if the mirror neurons' owners were experiencing the activity (you can see why they use the word). When you react to pictures of people getting flu shots, for example, your mirror neurons will act as if you had gotten the shot too. You wince as a result. We can detect both the reaction and the wincing with brain-imaging technology—very useful for looking at the neural circuitry of both cognitive empathy (Theory of Mind) and affective empathy (the feels).

A group of researchers from Canada used mirror neuron reactivity to observe what power does to our brains, designing an experiment similar to Galinsky's reminiscence-based efforts. Some subjects were asked to write an essay, recalling a time when someone had exerted power over someone else. Then the subjects' brains were examined using tests designed to assess mirror neuron activity. Their mirror systems sprang to life, scoring about a 30 in this lab (which is high).

Others were asked to write an essay recalling a time *they* felt power over someone else. Their brain activities were also assessed, just like the first group. Their mirrors didn't spring to life at all. In fact, the activity of these circuits fell below baseline. Average score around a *negative* 5 (that's low).

Just like the behavioral work previously mentioned, power could exert a force, even with something as fragile as reminiscence.

This study was only one of many showing what behaviorists had demonstrated before: power affects people's ability to respond to their world. Only, this time, the story was fleshed out in neurons. Mirror neurons predictably measure one's ability to understand the experience of others. Power, just as predictably, wipes it out.

What are the consequences of turning off your relational radars? Is it a bad thing to become socially blind as you begin to lead? The answer is yes, a depressing affirmation that can be measured in a variety of ways. The willingness to objectify and the ability to feel immune to correction and oversight are two that come to mind. We can illustrate both by examining a movie about band practice.

Instruments and instrumentality

The movie I'm describing is *Whiplash*, a 2014 release that takes place in a hyper-competitive, barely fictional music conservatory. The film garnered actor J. K. Simmons an Oscar and, man, did he deserve it. He plays a volatile jazz professor named Terence Fletcher, who is weaponized with an imperious teaching style to the right of Attila the Hun. He yells at his students, makes fun of their weight, intimidates and humiliates them, and even throws a chair at one, who responds by practicing so hard his fingers turn bloody. It's his band, Professor Fletcher tells them, and no incompetent kid will sully his reputation.

Fletcher's actions take their toll on the kids. Most are cowed into submission, transformed into objects for the pure goal of preserving a professorial monster's reputation. There are many reasons for the students' horror. The biggest being Fletcher's uncanny ability to objectify his students, one of power's most disabling effects on the human experience.

What do we mean by objectification? The definition I like—mostly because it can be measured—is the word *instrumentality*, which is the willingness to turn people into instruments useful for a leader's goal. Subordinates, stripped of their innate personhood, simply become means to an end. Research reveals that when people acquire power, they become willing to treat subordinates more like instruments, less like people. Throughout *Whiplash*, students were kicked out when they played out of tune, and replaced when they couldn't play fast enough. No regard was paid to them as human beings. The students were there only as instruments—both literally and figuratively—in order to make Fletcher's band sound good.

Though it's one of the great lessons of the Stanford Prison Experiment, the tendency toward instrumentality exists everywhere people obtain power: on sports teams, in business settings, even at band practices. One remarkable paper described instrumentality in six separate workplace-related experiments. These ranged from simply measuring the behaviors of executives versus subordinates to stimulating powerful feelings in otherwise power-neutral people. The paper concluded that "across the studies, power led to objectification, defined as the tendency to approach social targets based more on how useful they are than on the value of their less useful attributes."

To make matters worse, people in power feel free to objectify because they feel increasingly untethered from normal rules of behavior. The habit is common enough to have its own name: *hubris syndrome*. Empowered people increasingly feel entitled, as if they don't have to play by the rules.

How do we know this? Impunity attitudes were first shown in those power-primed experiments discussed earlier. Empowered people are 20% more likely to cheat under laboratory conditions (testing financial rewards in a lottery). Empowered people are also more likely to fudge on their taxes, keep a stolen bicycle, and blast through the speed limit.

This entitlement extends even into social relationships. As we will examine later, powerful people are much more likely to commit adultery, for example, and more likely to engage in unprotected sex. Researchers wrote about it this way:

... people with power not only take what they want
because they can do so unpunished, but also because
they intuitively feel they are entitled to do so.

A band leader who wins competitions, for instance, can justify abusing his pupils and can do so with a disturbing amount of impunity. He is, after all, the leader. The opposite, interestingly, is also true. Those people without much power often don't take advantage of the little control they have. They don't feel entitled to it.

Whether they are playing the right notes or not, subordinates lose when people with power believe they are entitled to wield it with impunity.

The maelstrom

As we have discussed, giving people power—even letting them conjure a memory of when they experienced power—is enough to change behavior. How do all these findings affect the workplace?

An executive is in danger of experiencing a behavioral perfect storm when they acquire new authority. If care isn't taken, that storm ends up creating a massive behavioral whirlpool, something the Dutch call a *maelstrom*. Newly empowered people are in danger of swirling down the relational drain.

You might have seen this scenario play out at a company you worked at: Someone obtains power, a promotion, for instance. That person's self-interest becomes heightened. Suddenly, that person might be more inclined to cut ethical corners. Their focus becomes less on how the group is doing and more on how they themselves are doing. People around them may begin to chafe, mixtures of annoyance and occasionally envy.

As this selfish vortex forms, the empowered person becomes increasingly tone-deaf to that chafing. They become unable to read a subordinate's face accurately. They lose sight of others' psychological interiors. Subordinates may become irritated at an executive's newly revealed selfishness, at the very time the executive is losing the ability to detect it.

The heartbreak in all of this is that a subordinate's feelings may not matter to their newly minted boss. Remember that power increases a person's willingness to objectify colleagues, to see them as instruments of their will rather than as people. It's a cruel switch. A newly empowered person loses the ability to empathize as well as the ability to understand. Given time, executives lose friends in the workplace. The best they can muster are allies.

To make matters worse, a creepy sense of entitlement takes over at the same time their emotional doppler is switching off. They've been tricked-out with an executive scepter; rules that apply to under-lings don't apply to executive aristocracy. And to top it all off, what do businesses typically give as a reward to the people they promote? More money—yet another way to give the newly promoted person even more ability and permission to do whatever the hell they want.

As bad as this behavioral weather is, we haven't even touched on the most socially turbulent consequence of this maelstrom. We are talking about sexual misconduct in the workplace. This consequence packs damaging winds so powerful that a national movement coalesced to combat it. Its origin, as we will see, stems from power. We will begin to unpack this tumultuous consequence with a few quotes from a few people who know quite a bit about power.

Singer, writer, conqueror, secretary of state

He was twice voted the sexiest politician alive, both recently and back in the day. The award was given to former Secretary of State Henry Kissinger.

Thick glasses, thicker German accent, droll public demeanor—sex is probably *not* the first thing you think of when you think of Kissinger. Yet his personal life was tabloid fodder for years. When asked why he waited until he was older to finally remarry (he'd been divorced), Kissinger channeled Napoleon Bonaparte, who supposedly said, "Power is the ultimate aphrodisiac."

As a woman coming up in the music industry in the 2000s, Janelle Monáe knows a thing or two about sex and power. In a song called "Screwed", Monáe sings a line that is often attributed to the writer Oscar Wilde, though its origin is unclear: "Everything is sex. Except sex, which is power."

Neither Monáe nor Kissinger might know this, but there's a fair amount of brain science to back them up. Findings fly in from numerous directions, both behavioral and biochemical, and from several continents.

We begin with a Florida State research study that enrolled both male and female participants. Using decision-making simulations and word-stem associations ("what does this word trigger in you?"), researchers tested the hypothesis that giving someone power activates a "mating motive." That is a Darwinian-enrobed euphemism for sexual arousal. Experiment subjects became more likely to sexualize their relationships by about 33% on average. Scientists noted that "Having power over a member of the opposite sex activated sexual concepts ... indicating the activation of a mating goal." They went on to conclude " ... it was power ... that caused an increase in sexual motivation."

The researchers named this idea *sexual overperception*. They also found that power didn't just increase the appetitive components of sex. Power also created some delusional aspects associated with sex, making people believe they were more sexually attractive than their subordinates judged them to be. This concept is called *self-perceived mating value*, and the feelings it engenders, *sexual expectation*. The rationale of people with supersized sexual expectations is delusional. They've suddenly somehow become so sexually attractive that, of course, subordinates are attracted to them. Interestingly, researchers found that power induced these feelings in both men and women.

We've seen the relational car wrecks caused by people in power due to the strong cocktail of sexual overperception (that increase in sexual appetite) and sexual expectation (that belief that oneself is more sexually attractive than peers believe). A 2018 *New York Times* article showed just how widespread the intoxicating addiction to power can become. It cited over 200 people—mostly males—who were relieved of their authority via resignation, termination, or actual arrest, solely due to reports of sexual misconduct. The #MeToo movement highlights just how much of an equal opportunity offender power can be: resignations, firings, and arrests have occurred from boardrooms to classrooms, from executive suites in Hollywood to congressional chambers in Washington, DC.

Clearly, sex and power make destructive drinking buddies. It's such a toxic dynamic—and, these days, so well defined—it warrants an examination on a biochemical level.

Blame it on the hormones

The behavioral research linking power to sexual overperception is robust. What about the underlying biochemistry? Does that change too?

Turns out, it does. We shift our focus now to the endocrine system, the network of glands in your body responsible for producing and secreting hormones. We will specifically examine what may be one of the most misunderstood of all human hormones: testosterone.

As you know, testosterone has been traditionally considered the "male steroid." It's the manly, guy-only, chest-beating origin of a thousand low-T television commercials. This drum-pounding isn't entirely accurate. Testosterone, it turns out, is not just for men. Years ago, the steroid was found to exist in measurable levels in both sexes (though menstrual cycles make its role in women a bit more complicated). When people become sexually aroused—that's the appetitive aspect of sexual feeling—testosterone levels increase. This is true for all of us.

Psychologist Cordelia Fine argues in her book *Testosterone Rex* that the male-only myth is just one of many myths surrounding the steroid. She also explains that the relationship between testosterone and specific behaviors—think aggression and arousal—is a gross oversimplification for either sex. It turns out that our bodies release hormones in collaboration with social frameworks, reacting to and exerting control over social situations.

Let's look at a male-only athletic example. One research effort from the University of Cambridge in England involved deliberately rigging a rowing contest with a bunch of male rowers. The scientists convinced some of these athletes they had won a rowing competition (even though they hadn't). Didn't matter: Their testosterone levels shot up by 14.5% compared with controls, who were told they had lost. The testosterone levels of the controls subsequently dropped by more than 7%!

Behavioral measures taken alongside the biochemistry metrics showed the effects this elevated testosterone has on arousal. Winning males, the researchers discovered, were more likely to approach a female to instigate casual sex. These males also displayed increased self-perceived mating value (sound familiar?). Quoth the scientists:

> ... the men ... (were) more likely to approach attractive
> women in an effort to instigate sexual relations.

What's more, the research confirmed what Fine argued in her book. Researchers found that "the endocrine system that controls hormones is responsive to situational changes."

That last quote is very important. The acquisition of power is just the type of situational change that kicks our endocrine system into high gear, dispensing hormones such as testosterone.

Why does power do such extraordinary things to ordinary people, to the point of altering our biochemistry? There are no easy explanations, but the guess is that it's a product of multiple factors—a mixture of our evolutionary past, including our need to be so social with each other, and our brain's obsession with energy conservation.

Friends and energies

No question we're a social species, relational as greeting cards, and not by accident. As we have discussed, being a social species was vital to our survival from an evolutionary standpoint. So, why does power so capably degrade Theory of Mind and empathy in people with power? The answer is humiliating: it's because we act a lot like apes.

Researchers discovered long ago that we create social hierarchies just like many other primates. We just took our preoccupation with these hierarchies further than our genetic cousins did, and as a result, ours are more sophisticated. We spend a tremendous amount of time wondering how others view us, how we might predict others' behaviors, and maybe how we can manipulate them. We fashioned the concept of ally. And enemy.

Such sociality is very expensive bioenergetically. The brain already gobbles up 20% of the energy we consume (though it occupies just 2% of our body weight). Much of this energy is diverted to establishing and maintaining social relationships. It's one reason why people often say they're exhausted after a social engagement, such as a party.

Why were we willing to pay such a price? The answer is once again humiliating. Social relationships are so important because we were—and are—biological wimps. From our tiny teeth to our fragile fingers, we're ill-equipped to fight most of the big creatures on the planet. It's enough to make you wonder how we became the world's apex predator.

But apex predator we did become. Our energy-saturated brains spent the overflow learning to cooperate. Theory of Mind and its first cousin empathy probably did most of the heavy lifting. After all, if you can predict somebody's intentions, you can also forecast how they'll react under specific circumstances, and empathy will allow you kindness while you're busy prognosticating.

That's a remarkably useful talent in the friendless world of the Serengeti. Coordinating hunting comes to mind. So does keeping track of children. You essentially double your biomass, not by actually doubling your biomass but by establishing a confederation with someone else, maybe even a friendship. The idea is formally enshrined in something called the *social brain hypothesis*.

How does this all connect to powerful people becoming less relationally competent? There are probably several reasons, including the famous "it's lonely at the top" perception. Certainly, newly minted leaders face whole new sets of relational challenges. People start to be nice to them, seeking not friendship but favor. If that happens regularly, as it often does, a leader may suspect everybody has an ulterior motive. They then start feeling isolated, interacting less, and losing social skills because of a lack of practice.

But that's not the whole story. Other investigations suggest that loneliness at the top may be less of an issue than commonly thought. One group of researchers, looking at loneliness and executive authority, expressed it this way:

> *We speculate that the psychological benefits of power can substitute for the human need to belong to social groups.*

In other words, if you use alliances for survival, once survival isn't an issue, you won't need allies as much. Human psychology adjusts accordingly. This group of researchers go on to explain, "The results were clear: power decreased loneliness by reducing the felt need for affiliation with others."

Examine this from an evolutionary standpoint: once power ensures you don't need other people for your survival, the brain's negative incentive to get you to be more social, loneliness, isn't as useful. The positive tools of coalition-building—such as Theory of Mind and empathy—aren't as essential either. Why burn excess energy maintaining alliances if you don't need them? In the heart of Darwinian darkness, friends don't serve quite as useful a purpose to those with power. It's draconian, critical, and predictable, but in the harsh world of the Serengeti, species always hand the mic over to survival.

Coping with layoffs

There might be another reason why power and empathy are inversely related. It has to do with an activity many executives state is the hardest part of their job: letting people go.

When people exert the levers of power in situations where people will be fired, internalized conflict is a common first reaction. Executives are seldom immune to "blood on their hands" feelings, at least at first. Many begin struggling with several of the bellwethers of depression: loss of sleep and stress-related health problems, for example. To keep executives from being overwhelmed by their responsibilities, coping mechanisms have to kick in.

One common way of coping is tactical retreat. Executives some-times begin emotionally withdrawing from colleagues who must suffer the consequences of their authority. Or they do what the mili-tary trains its soldiers to do: begin thinking of people not as humans but as targets. Supervisory language might become technical and distant, conversations increasingly filled with euphemisms ("we're moving in another direction"), as if that might take the edge off the sharp ax they are compelled to wield.

It works. After a period of time, the empathy switch turns to the off position. That's not surprising, given that power already coaxes people to objectify their relational surroundings. The extreme form of this withdrawal is called *moral disengagement*. The executive shel-ters in place emotionally. Which may explain the social bluntness so common with empowered people.

Executives become increasingly emotionally incompetent, allowing them increasing emotional protection. Mild delusion may set in. They might downplay the consequences of their actions, thinking that if they can't perceive the damage they're doing, maybe they're not even doing it.

From a brain science perspective, the whole thing smells like energy budgeting. Which is what it is. This happens all over the world. Most of us researchers think it has happened for eons, the biggest price humans pay for being able to dominate the world—and each other.

Hardened hierarchies

Energy bookkeeping might explain why power leads to a loss of relational skills, but what about power's effect on sex? Are there Darwinian reasons why power keeps a supply of Viagra in its behavioral tool kit?

There are indeed, say most evolutionary biologists. These reasons have to do with how most social mammals survived without needing to become the size of mastodons (that biomass-doubling thing). It's axiomatic, according to these scientists, that dominant alpha primates get all the sex. Makes sense, because then they also get all the progeny, which creates a more powerful home team with better chances for survival.

Is that also true for humans? The answer is a qualified maybe.

Social structures are more complex in humans, so the correspondence isn't one-to-one. Robert Sapolsky, neuroscientist at Stanford, rightly relates that a person can be low status in one social circle but completely dominant in another. This complexity has not, however, divorced us entirely from our paleontology.

Most researchers feel that ancestors common to both chimps and humans (we diverged between six and nine million years ago) had alpha characters. Many of them believe this unequal behavioral structure cast long shadows clear into the twenty-first century. The research we've covered suggests there may be some truth to this.

For example, as soon as you acquire power, one way to ensure survival is to become more willing to sexualize your relational world.

Thus, you become more likely to be aroused when relating to others. It's power first and, with lust following, super-babies afterward.

Testable hypotheses come from these ideas. We've already discussed two of them: sexual overperception (the more power you accumulate, the greater your sexual interest becomes) and self-perceived mating value (the more power you gather, the more sexually attractive you think you are). These behaviors don't happen in evolutionary vacuums. All are tilted toward increasing the probability of giving us humans, a weak species in the wild—which, by the way, is fertile only a few days a month—the best shot at survival. The fact that we no longer live in a savannah hasn't convinced our brains to disabuse itself of this fact.

Prophylactic education

Is there anything that *can* convince our brains not to fall prey to the pitfalls of power?

The answer is yes. Happily, the behavioral sciences do have something to say about it. The answer, unlike so many translational research efforts, is actually pretty simple: warn people.

Before new executives take their new jobs and first grip the levers of power, they need someone to sit down with them and give them a warning. They need to hear in advance what power is statistically likely to do to them and their relationships with subordinates. They need to know what their vulnerabilities are likely to be. They need to know how entitlement and impunity accelerate, how sexual over perception occurs. New executives should examine the data behind this chapter or, more simply, just read the whole thing.

There's an empirical reason behind creating such a come-to-Jesus moment. Whether you're about to promote a colleague or experience a promotion yourself, simply knowing about power's morally icky and potentially costly temptations can be an effective talisman against them. Researchers even have a term for this type of prevention. They call it *prophylactic education*.

Not convinced it's that simple? Prophylactic education did wonders in the medical profession for combating malpractice. Researchers found that the most common malpractice complaints stemmed from communication issues before a surgical procedure ever took place. These complaints were basically variations of "the physician did not warn me what might happen in advance."

Researchers got a bright idea. What if physicians did warn patients in advance? What if they pre-armed their patients with knowledge? If the doctors made a mistake, what if they told patients about the error, rather than hide it? What if they even *apologized*?

The results were astonishing.

Patients given such pre-operative coaching about their surgery experienced reduced anxiety and depression in post-op, used fewer painkillers, had fewer complications and, to the delight of administrators, experienced shorter hospital stays. At the University of Michigan, the number of lawsuits dropped by almost two-thirds. Administrators spent 61% less on lawyers. Later audits showed a 58% decrease in the rate of claims (per patient), even though the hospital's medical activity climbed 72% in the same period. More activity occurred, and it cost less.

The bottom line? It turns out, prepping people in advance is good business. Researchers even discovered why it works. When similar programs were adopted at the University of Heidelberg, the researchers found that "... education and support had a positive effect on surgical patients' physical and psychological well-being before and after surgery because education and support maintain or increase the patient's feelings of control." In other words, prophylactic education works because it gives people the power of prediction, much like severe-weather forecasts allow people to get ready for coming storms.

But hospital rooms aren't the same thing as C-suites. Would giving new execs knowledge about what was likely to happen to them when they were handed power create similar feelings of control? Might prophylactic education work in business?

Even though medicine and business are far apart professionally, it doesn't seem to matter. The answer, happily, is yes.

Hard changes

Behaviors don't just trim their sails and sail away because you ask them to. Sadly, successful change is rare. But it's not extinct.

The few programs capable of preventing the collateral damage of power mostly involve knowledge transfer. One interesting effort involved studying how to keep male-female relationships professional, especially with an asymmetric distribution of power. This is the research of psychologist David Smith and sociologist Brad Johnson. They explored that trickiest of professional interactions: mentoring—specifically *men* mentoring *women*.

They found that mentor relationships were most consistently kept on professional headings if the participants knew in advance the behavioral pitfalls that could occur. They discovered that if men and women knew something about the behavioral science behind attractions—sexual overperception and emotional vulnerability—they were less likely to engage in inappropriate workplace relationships. Smith and Johnson published their findings in the *Harvard Business Review*, then detailed them in their book *Athena Rising*.

Educating people about the underpinnings of their behavior *works*. Such prophylactic knowledge is powerful enough to keep the relational tones friendly but professional. It's very much in line with what physicians did to reduce lawsuits. That's why the best solution is to tell people who've been promoted what's likely to come.

I must mention that confirmatory studies are not a deep bench here. This type of behavioral work is woefully underfunded. I should also mention that, while education is powerful, it's not all-powerful. If an audience doesn't believe the peer-reviewed data, exposing them to such facts simply hardens their disbelief. It's just more "fake news" to them. Only when audiences assume credibility in science does telling them about mechanisms become a change agent.

If employees, executives, and companies take these data seriously, bolting early warning content into their management programs, they're likely to save a lot of money and, as the beliefs and behaviors of executives in power change, a lot of heartache too. These data are potent enough to hold at bay the monsters that power can create, ensuring that those in authority don't aim to become the Lord of All the Beasts. Or the King of Scotland.

POWER

Brain Rule: *Power is like fire. It can cook your food or burn your house down.*

- Power is the ability to alter another's condition or state of mind by manipulating resources and punishments.
- Having power can cause you to place your own interests above those of a given group, reduce your ability to detect emotions and empathize, increase your sexual appetite, delude you into thinking you're more sexually attractive to others than you actually are, and even lead you to compromise your ethics.
- High-net-worthers are more likely to lie during negotiations, cheat when they believe they could win a prize, and take goods for themselves even when they are valuable to others.
- To safeguard against the harmful qualities of power, learn about them and temper your expectations accordingly.
- Prepare employees who are about to acquire power (via promotions and raises) by warning them about the potential pitfalls of power. Doing so makes them less likely to fall victim to those pitfalls.

presentations

Capture your audience's emotion, and you will have their attention (at least for ten minutes).

WHO WOULD'VE THOUGHT that a TV commercial that contained soda pop, cheesy '70s disco music, and a football jersey would be one of the most iconic advertisements of all time?

The commercial begins in the tunnel of a stadium. To the right is a little nine-year-old boy, holding what's supposed to be the star of the commercial, a bottle of Coke. But the real celebrity is to the left. It's Mean Joe Greene, a football player seemingly born with his shoulder pads on, one of the most ferocious and feared players of 1979. He's not very scary at just this moment. Sadly, he's been injured and is limping down the hallway, head bent, jersey ripped off and draped over one shoulder.

"Mr. Greene? Mr. Greene?" the boy calls. Greene pauses, annoyed. "Yeah?" he snarls back.

"Ya need any help?" the boy innocently asks. Greene refuses and continues hobbling down the corridor. The boy, undeterred, declares, "I just want you to know I think you're the best ever!" And then, "Want my Coke? It's okay. You can have it."

The reluctant giant pauses and seems to soften as the kid hands him the bottle. Greene takes an NFL-sized guzzle, the soft drink's logo prominently displayed. The disco music is turned up while Greene drains the bottle dry. The little boy sighs while this is going on, looking ignored and a bit crestfallen, and turns to leave.

"Hey kid," Joe says, no longer grimacing. "Catch!" He tosses him his jersey, eliciting a playoff-sized smile, winning the hearts and minds of just about everybody who watched it, including professional advertising execs. It won most major advertising awards in 1979, including the prestigious Clio and the Cannes Gold Lion, and was imitated around the world. *TV Guide* listed it as one of the greatest commercials of all time.

That's no accident.

There are elements in this ad, placed deliberately, that are capable of capturing our attention and gluing themselves to our neurons. We're going to explore the ingredients of this adhesive. We'll find these elements useful for holding the brain's attention not only in a TV room but also in a boardroom—or in a classroom, or in any room where it's important to get people to listen to what you have to say.

Spotlight

So, why do we remember some pieces of information but not others? The obvious answer is that we pay closer attention to objects we find compelling, and they become more memorable because we pay closer attention to them.

While that bit of near-circular reasoning is mostly true, it's also bewildering. There's no shortage of competing hypotheses to explain why we pay attention to certain details and put others immediately out of our minds.

The first is called *attentional spotlight theory*. The idea originally came from vision research, with scientists asking what causes our eyes to land on some things and not others. Scientists such as Michael Posner quickly realized that was the wrong question. He found that our eyes didn't need to land on something for us to pay attention to it. (Attention, it turned out, was not gaze-dependent.) Instead, there appeared to be an anxious steering committee inside our heads, scanning various sensory experiences, such as the feel of a cold breeze or the smell of a fire burning, seeking something interesting upon which to focus. This steering committee then makes a handshake agreement about what we should be paying attention to, an agreement based on two inputs: what our brain thinks it should be paying attention to and what is actually out there. Researchers even thought they discovered where this nervous committee lived, in the prefrontal cortex just behind your forehead.

Not everybody buys into attentional spotlight theory. Many believe it doesn't take into account everything known about how we pay attention. They point out that the theory has a hard time integrating what researchers call a *filtering system*. Researchers critical of the theory think we pay attention to some inputs at the expense of other inputs. At the very least, our sophisticated attentional steering committee consults an equally sophisticated don't-focus-on-this-input subcommittee during processing. This subcommittee is not in the prefrontal cortex, says the anti-spotlight crowd. They point instead to a region called the thalamus, an ancient neural structure deep inside our heads that normally functions as a kind of air traffic control system for motor and sensory signals, alerting specific regions of the brain to process specific types of input. There is evidence the thalamus also supervises attentional filtering functions (at least if you're a mouse).

The jury will have to await further inputs before we finally know what neurological bureaucracies supervise attention. Fortunately, we don't need a complete picture to understand where our attention goes during presentations. Clearly, the organ finds some inputs interesting, other inputs boring. Let's look at some techniques that will help us avoid becoming a verbal sleeping pill when people start listening to what we have to say.

Minutes

So, what goes on in our brains when someone fires up the PowerPoint deck and begins talking? How long will we pay attention to the speaker before we start to check out?

The research literature here is a bit mixed. There's some evidence that attention can lapse as early as thirty seconds into a talk, which has given birth to the anecdotal "if you don't do something in the first half-minute, you'll lose your audience" mantra in public speaking courses. The direct evidence for this is fairly weak, but some studies indirectly show how one could come to this captivate your audience in thirty seconds maxim.

These findings are assisted by the robust and somewhat disturbing literature on first impressions. Research shows that we make remarkably durable judgments about people only a few moments after meeting them. Within 100 milliseconds (a tenth of a second), we have already assessed a person's likability, trustworthiness, and competence.

Is that speed relevant to presenters? Though direct evidence is sparse, it's hard to believe these 100-millisecond, rush-to-judgment detectors would remain silent, so it's safe to assume the initial few minutes of presentations are important, especially if there are strangers in the room. If you're giving a talk, the first words out of your mouth should probably be memorized ones.

But that's not the only critical time point. Research shows that people try valiantly to stay up with even boring speakers, attention waxing and waning—until about ten minutes into the talk. Then something happens. A law of speaking sometimes called "the ten-minute rule" takes effect. Psychologist Wilbert McKeachie found, when it comes to listening to a speaker talk, the attention of the audience falls off the cliff after about ten minutes. If you don't mount a rescue operation by the nine-minutes-and-fifty-nine-seconds mark, you'll lose the room.

Confirmation of this ten-minute rule has come relatively recently. Published in the science journal *Nature*, Robert Ewer found that audiences will stay with a speaker for about ten minutes (Ewer's exact time was eleven minutes, forty-two seconds). But if the speaker doesn't do or say something that recaptures the audience's interest at that mark, attention flags, just like McKeachie found. If the audience still finds the speaker boring at thirteen minutes, twelve seconds, the speaker should probably just leave the podium. Ewer even calculated the rate of the decline after that critical point:

For every seventy seconds that a speaker droned on,
the odds that their talk had been 'boring' doubled.

Yep. *Doubled.*

Emotions

The data suggests you really need to do something to staunch the attentional bleeding around ten minutes or risk losing your audience. But what type of tourniquet works? Researchers have a clue. It involves solving a big, fat problem.

The big, fat problem concerns the amount of information being shoveled into the brain at any one time, whether sitting in a conference room or simply lying in bed. Research shows it's almost always too much. Take the inputs from just one source, visual information. Neuroscientist Marcus Raichle explains that the equivalent of about 10 billion bits of visual information per second slam into your retina the instant you open your eyes. But the retina, overwhelmed as an emergency room, can only process about 6 million of those bits at a time. And further in the back of your brain, where visual perception begins occurring, that number shrinks to about 10,000.

That's obviously a bottleneck problem, and we've only discussed one sense. The brain has to process not only external sensory inputs from at least four other sources but also internal inputs, such as positional information supplied by your inner ear. And hunger information supplied by your stomach. And information from every other part of your body. If the brain didn't deploy some type of filtering system, one capable of prioritizing inputs into a let's-do-this-first list, then its networks would be constantly suffering a mass-distributed denial of service. You wouldn't perceive anything at all.

Fortunately, the brain has a defense against overload, one capable of creating orderly priorities amongst the crowded information stream. That's what we think emotions do. Even though they have a reputation for being messy, emotions actually provide the organizational prioritizing necessary for us to pay attention to some inputs at the expense of others. Indeed, the more emotional a stimulus becomes, the more likely we are to pay attention to and remember it. Researchers give such powerful attention-grabbing inputs a name. They're called *emotionally competent stimuli,* or ECS.

Exactly what types of stimuli elicit the strongest attentional responses? Two broad categories crowd the waiting room, all traceable to evolutionary urgency. We pay tons of attention to threat, for example. That's because evolution honed our concern with survival in the present. We also pay tons of attention to sex. That's because evolution honed our concern with survival in the future. Indeed, projecting one's genes to the next generation is the whole *point* of evolution. Emotions force our messy inputs to line up according to biological priorities.

As you have probably already guessed, an ECS has a big say in what you should being doing nine minutes and fifty-nine seconds into your talk.

Hooks aren't just for fish

I call ECS "hooks," but they're really just simple strategies to freshen the audience's attention. Around the ten-minute mark, you need to give your audience a compelling reason to listen for another ten minutes. You need to give them a hook. Then, ten minutes later, you need to give them another. This carries on throughout the life of your presentation.

Here's an example of a hook I use when lecturing about the social development of young children. It's a story from the Art Linkletter segment on the show *House Party* called, "Kids Say The Darndest Things, an afternoon family program from the 1960s. Linkletter often interviewed children, asking them open-ended questions. Their answers were sometimes quite revealing—sometimes *too* revealing. You should know every show was a live television experience in those days, broadcast in the days before videotape.

During one broadcast, Linkletter asked a little boy named Tommy what would make him happy. Tommy answered, "A bed of my own. That would make me happy."

Linkletter grew concerned. "Don't you sleep in a bed?"

"I usually sleep with my mom and dad," Tommy responded, "but when my dad is gone, mom sleeps with Uncle Bob, and I have to sleep on the couch. And anyway, he's not really my uncle."

I can just hear the director in the control room shouting, "Cut to commercial!"

In my experience, the best hooks possess four characteristics:

1. Emotionality

First—and this is not optional—the hook must be *emotional*, capable of activating the ever-vigilant attention mechanisms in your listeners' brains. Appeals to threat and survival are powerful; it's the reason why I opened this chapter with the story about the magic between an injured football player and a little boy. Hooks could also be about sex, though the emphasis should be on the results of reproduction—babies and puppies come to mind—rather than on the act itself (which might not elicit the response you're looking for). Humor works here too. Like sex, laughter gives the brain a shot of dopamine, the feel-good hormone brains usually use to reward themselves.

2. Relevance

The hooks should be relevant to the material at hand. You could simply crack any old joke to get your audience's attention, but presentations aren't usually comedy hours, and presenters aren't usually stand-up comedians. Make sure your hook either summarizes something you just said, illustrates something you're currently saying, or foreshadows something you're about to say. Educated audiences can understand the difference between presenters trying earnestly to communicate important information and presenters simply trying to entertain them. To that point, hooks should also fit the emotional tone of the presentation. Bad jokes will be memorable for all the wrong reasons.

3. *Brevity*

Your hook should be short. Many hooks are powerful enough to overwhelm your content. It can lead your audiences to remember the ECS rather than the nine minutes and fifty-nine seconds of content preceding it. You can solve your emphasis problem by simply limiting the amount of airtime you give a hook. In my nearly forty years of lecturing and teaching, I've found that a hook lasting about two minutes is just about right.

4. *Storytelling*

Whenever you can, transform your hook into a narrative. Stories are some of the most memorable, powerful ways to continuously command attention.

The Art Linkletter story displays all four characteristics of a competent hook. The sex-and-humor sections obviously fulfill the emotionality requirement. Because the keynote is about the developing social awareness of children (I usually use it to open such lectures), the relevancy component is checked off. It's short, too. I can usually describe the incident in less than a couple of minutes. And, above all, it is a story.

Stories play such a powerful role in the human experience that I'm about to devote the next few sections to them.

The components of a narrative

We'll start with a definition of a narrative (or story—I'll use them interchangeably here) and compare it to a plot. This task is more difficult than one might think, regardless of the term used. Fortunately, we can recruit some professional help from people who write stories for a living: writers of fiction. We'll begin with the celebrated novelist E. M. Forster. He thought about storytelling so much, he wrote a book about it. In those pages, he famously compared these two sentences:

1. The king died and then the queen died.
2. The king died and then the queen died *of grief.*

These sentences do not occupy the same emotional universe. Nor the same definitional one. Forster says the first one's a story, the second a plot. The difference? The first sentence simply describes the facts as a newspaper journalist might write them: this happened, then this happened. The second sentence describes the relational gravity that binds two people together, an emotional galaxy revealed in just two words. That deep view makes all the difference.

According to writer Janet Burroway, "a story is a series of events recorded in their chronological order. A plot is a series of events deliberately arranged so as to reveal their dramatic, thematic, and emotional significance."

Academics have spent decades trying to understand exactly what turns stories into plots. They've concluded that a plot is a story that has a dramatic structure. So the next question becomes: Of what does that structure consist?

Confusion reigns once again. Literary theorists believe basic structural elements exist in all plots, but nobody agrees exactly what they are, or even how many there are. You can find papers declaring there are seven elements, or thirty-one elements, or twenty. A playwright named Gustav Freytag devised a dramatic structure with five acts that unfold over time—essentially a tension-and-release model—which, for some strange reason, is called *Freytag's pyramid* (though it looks more like an inverted square root).

Modern behaviorists haven't fared much better in coming up with a formula that nails down narrative. Some define narrative as the interaction of "intentional agents," redolent of Theory of Mind. Some characterize narratives as causally linked events sporting a time stamp, something like our queen's death. Some neuroscientists think narrative processing involves a cognitive gadget called *episodic memory.* This gadget works like a film editor, dividing specific experiences into segments that are more easily storable.

These efforts leave out many details that may prove crucial to the concept. Yet narratives—whatever they turn out to be—don't seem to be ephemeral. We even think we know where narratives are processed in the brain, depending upon how you define them.

When it does finally come down to the definition, I have to throw up my hands and surrender to yet another writer, one of the greatest storytellers I know: Ira Glass. He's a famous radio personality and host of the long-running, award-winning *This American Life.* Glass says narrative is "like being on a train that has a destination" and that, at the end, "you're going to find something."

Until science can come up with a more precise definition for a narrative, pin down its inner workings, and map out its mechanics, this beautiful little idea will have to suffice, even for a fussy scientist like me.

Narratives and attention

What happens when the brain detects a narrative? Research shows that the organ lights up like a pinball machine, and in a coordinated fashion. For years, researchers have been trying to discern exactly which areas of the brain activate when it experiences a narrative.

One well-established finding concerns the activation of the brain's attentional systems. They're put on high alert when the organ thinks narratives are forming. (Controls consisted of exposing subjects to a mental-math task, which didn't result in alerts at all.) This may explain why narratives make powerful attentional hooks during presentations—and statistics don't.

But attentional networks aren't the only regions stimulated by narratives. Language areas experience surges in electrical activity too. So do motor areas, especially those regions mimicking the action content of the narrative being experienced. The areas in charge of interpreting your sense of touch, sight, and smell do too. (Just reading the word *cinnamon* stimulates the brain's regions that decipher olfactory signals, for example.) The brain begins imitating elements of the story as soon as its tissues absorb it.

Scientists call such mimicry *narrative transportation*. It's when our brains buy a first-class ticket to anywhere a good book tells us we should go, all for the simple price of reading it. We're fooled into thinking the train is really going somewhere, even though the train only consists of words on pages. It's probably the reason why we often imagine being in the place we're reading about while we're reading about it. *Narrative transportation* is, if nothing else, an accurate term.

Another process stimulated in our brains as they experience narratives involves education. When researchers observe large neural networks lighting up, the memory scientists in the room take notice. The reason concerns a principle of information-processing they learned years ago. The more neural substrates stimulated at the moment of learning, the more robust the learning becomes. If you simply add an audio track to a visual presentation in a series of lectures, for example, the number of lectures required to master the content is reduced by 60%. Bonus points if you include other senses, such as olfactorial, gustatorial, or tactile. The reason multisensory experiences work so well is an access-point argument: the more regions stimulated at the moment of learning, the greater the number of access points created. These access points allow for easier retrieval later.

Which predicts something simple. If narrative stimulates so many areas in the brain at the moment of learning, shouldn't narrative also improve retention way after the moment has passed?

The answer's yes, though it's not a simple story.

Narrative improves retention

To help explain it all, I'd like to describe an incident that happened to me as a young teenager. I'd just finished reading *The Lord of the Rings* trilogy, and at the end, I cried. The images, lore, scholarship, and world-building were not things I'd previously encountered. In that moment, I remember praying to God that when I died, I didn't want to go to Heaven; I wanted to go to Middle Earth.

The images I conjured way back then remain with me to this day, as do their value. In fact, I cherish them so much, I've vowed never to see the movies—a promise I have kept, much to my family's chagrin. I simply couldn't bear to have director Peter Jackson's images interfere with the world inside my head. Those memories are precious and *indelible*.

Boosting recollection is exactly what narratives do to people. Stories not only grab our attention but also usually fasten a drop of superglue to make the information super-sticky. This adhesive property has been measured in labs from Stanford to New York City.

One Palo Alto experiment, famously done by brothers Chip and Dan Heath in a business class, involved students giving presentations. The students were assigned the task of convincing their colleagues that nonviolent crime either is or is not a problem, and they had one minute to do it. Retention tests followed. The Heaths found that the sixty-second presentations were saturated with statistics—2.5 on average. Only 10% of the presentations used narratives to persuade their audiences. When retention was assessed, only 5% of the class remembered any individual statistic, but 63% remembered the narratives.

Jerome Bruner, one of the titans in the field of developmental psychology, provided the New York point of view. Bruner was a cognitive psychologist with a broad range of interests, from baby brain development to education, focusing later in life on the role of narrative in cognition. He's been cited a lot, as has his research into memory and storytelling. Here's a quote from award-winning journalist Gaia Vince, citing Bruner's work:

> *Information told through stories is far more memorable—22 times more, according to one study—because multiple parts of the brain are activated for narratives.*

The interpreter

To what "multiple parts" was the above quote referring? Many brain scientists are working on understanding the neurological basis behind narrative processing. The fact that memory is so deeply involved gives them clues. To understand these clues, we need to discuss how brains manufacture different types of memories.

Notice I used the phrase "different types of memories," instead of the singular, all-encompassing word *memory*. That's because multiple memory systems exist in our brains, most of them operating in a semi-independent fashion. Surprised? Recalling that George Washington was president is processed in different regions than remembering that touching hot stoves will burn you. And both are processed in areas different from remembering how to ride a bicycle.

Which systems are involved in narrative processing? Researchers believe at least two types of memory systems are necessary. One is *semantic memory*, the memory for facts and concepts. Your ability to know the event you attended yesterday was your best friend's wedding—and that they served chocolate cake—are good examples of semantic memory. The other type, mentioned previously, is *episodic memory*. This is memory of events involving specific characters interacting through space and time. Your ability to recall who walked down the aisle first at your friend's wedding, when the preacher started his sermon, at what time the reception took place—these are all the domains of episodic memory.

These two systems give us clues about how the brain processes narratives, at least from a memory-systems perspective. They don't say much about where the actual narrative processing occurs, however. For that, we'll need researcher Michael Gazzaniga's insights. He's probably most famous for investigating cognitive functions that occur in specific sides of the brain, something called *functional lateralization*. Your ability to create and understand speech is located on your brain's left side, for example. Your ability to comprehend the emotional content of speech, in contrast, is located on the right.

Gazzaniga believes the brain possesses a narrative "factory," a story generator, and that it is similarly lateralized (mostly left side). He calls this factory "the interpreter." It has an aggregating function, combining small pieces of fact with large pieces of timing in order to manufacture a story.

Oddly enough, this includes the narrative that makes us who we are, our personal story. Though it sounds a bit like consciousness (whatever that is), many narrative processing features are rooted deeply in what makes "us" us. Science fiction author Ted Chiang renders this idea beautifully:

> *People are made of stories. Our memories are not the*
> *impartial accumulation of every second we've lived;*
> *they're the narrative that we assembled out*
> *of selected moments.*

Nobody teaches us how to construct these narratives, by the way. We seem to both form stories and rely on them automatically, perhaps inherently, outside our conscious awareness.

Evolutionary considerations

The brain's ability to automatically form and depend on stories is catnip to evolutionary theorists. It suggests the presence of selective, nonelective pressure on specific cognitive skills. But therein lies a puzzle: What possible survival advantage could the ability to create imaginary scenes, images, and characters around, say, the *Lord of the Rings* trilogy, confer on our illiterate, brutish brains?

Theorists have their explanations. One is the familiar efficiency argument. We've already observed how much brains love energy conservation. If narrative information is twenty-two times more memorable, then the brain pays a smaller retrieval bill than information twenty-two times less memorable. Ka-*ching* goes the bioenergetic cash register.

The second argument concerns nongenetic, intergenerational information transfer. Theorists believe ancient stories, probably like their modern counterparts, included illustrations of tribal institutional knowledge. Such knowledge might include social customs, such as pair-bonding rituals, or explanations about food foraging or hunt coordination or vanquishing enemies. All this knowledge could be transmitted to the next generation with a simple campfire story. That's especially useful when the alternative is waiting millions of years for DNA to do a similar, far less precise job.

Perhaps the most convincing reason the brain clung to narrative as a useful evolutionary mechanism concerns something we covered in chapter one. Remember Theory of Mind, that cognitive gadget allowing us to understand the intentions and motivations of others? One of the most intriguing characteristics of Theory of Mind is its ability to perceive something that cannot be easily apprehended with the five senses. Intentions don't have a visible body part, after all. It requires our imagination to perceive them. Understanding the main characters in a story does too: we envisage their perspective, an emotional version of narrative transportation. In competent authorial hands, this perspective-taking allows us to feel the impact of a character's past, make sense of their present behavior, imagine their future, and even forecast how we'd behave under similar circumstances.

I once got a lesson in the linkage between narrative and imagination while listening to an NPR interview with the author Robert Swartwood. He had challenged people to write a complete novel in

25 words or less, then gathered his favorites in an anthology called *Hint Fiction*. From the "Life and Death" section came this entry:

"Golden Years" (by Edith Pearlman)

She: Macular. He: Parkinson's. She pushing, he directing,
they get down the ramp, across the grass, through the
gate. The wheels roll riverwards.

This little snippet works because of the brain's ability to fill in the negative spaces described between the words. It is a powerful thing to weld Theory of Mind to our love affair with narratives.

Researchers believe this imaginary ability served a purpose beyond just thrilling fans of fiction. From an evolutionary perspective, it allowed us to *practice* interacting socially before actually *having to* interact socially. We could exercise our relational and cooperative skills—something we desperately needed in order to survive—without the real-world consequences of getting something wrong. Stories might have served as relational flight simulators, to paraphrase researcher Keith Oatley, allowing us to sharpen the critical skill of learning how to get along with each other.

If that's true, a judicious use of narratives after ten minutes of nonnarrative content is not only a good idea, it's pretty much a non-optional idea. It is difficult, after all, to fight with a preference millions of years in the making.

Dual-coding theory

Another issue millions of years in the making concerns that other sense pressed into service during presentations: eyeballs. Attentional research has a long history of looking at what others are looking at, then asking why they bother. It's even touched infant research. As you might recall from the leadership chapter, one powerful way to extract attentional information from babies involves observing their gaze, then measuring the length of stare. The longer they stare at something, the more they're presumably interested in it.

There's even an adult version of this correlation. Milgram published a paper in the late '60s, describing social interactions and staring. (To date, it remains the only research paper at which I laughed out loud while reading.) Milgram recruited actors to stand on a crowded street corner and simply look up, staring at a building window. He wanted to know if he could get strangers to stop what they were doing and stare at the same window. What he found was hysterical. People really did stop to look. The more actors you added to the corner, the more likely people were going to follow suit. Two actors got 50% of passersby to look up, and fifteen got 80%. (Repetitions of this famous experiment have found similar results, though the copying behavior is much weaker, and the effect didn't change quite so much when more actors were added.)

The important point is that where—and how—we use our eyes tells us something about what we're attending to, regardless of age.

We've already discussed how to attract and hold attention through content and structure, mostly describing verbal information. What we left out were visual experiences, which for most business professionals means slides. How do the visual processing centers of the brain deal with PowerPoint?

It's far beyond the scope of this book to unspool the science behind how brains process visual information that's presented in glowing chunks of 1280 x 720 pixels. Fortunately, we don't have to, because one influential idea called *dual-coding theory* seems almost tailor-made to take PowerPoints into consideration. It was championed by the late psychologist Allan Paivio, who also happened to be a bodybuilder.

Paivio's muscular research idea posited the existence of two large, independent paths for learning information. One path stored verbal inputs, something he called the "logogen pathway." The other stored visual inputs, something he called "imagen pathway." When listening to an audio input while looking at a PowerPoint presentation, the brain must immediately decide which of the two types of information it is perceiving. The organ then shuttles the information to the appropriate processing channel. If a person hears the word *spreadsheet,* for example, that information routes to the logogen pathway. But if the person sees a picture of a spreadsheet on the slide, that information routes to the imagen pathway.

The pictorial superiority effect

Paivio thinks these two pathways, though distinct, talk to each other like texting teenagers. The mechanism they employ is cellular in origin (nerves, not smartphones), using something called *referential interconnections.* This is where information stored in one pathway triggers similar information stored in another. Gazing at pictures of raptors might trigger the word *eagles,* for example, or *hawks* or *basketball teams,* the imagen store stimulating items in the logogen larder.

There's evidence imagen information is far better at creating referential connections than its logogen counterpart. That has direct implications for users of PowerPoint. Why? It helps explains something called the *pictorial superiority effect* (PSE).

The central tenet of PSE is that pictures are remembered better than words, in almost every way you can define "remembered better." It's been proven in item-recognition tests, paired-associate learning tests, serial recall/reconstruction tests, free recall tests—the list of obscure psychometric assays demonstrating PSE seems endless. The memories laid down by images are much more stable too. One experiment showed a PSE image stayed with its subjects *decades* after primary exposure to a specific stimulus. Pictures are also dealt with quickly. The brain can process an image even if the eye sees it for only thirteen milliseconds.

One effective example of the potency of PSE involved the 1964 presidential race between democrat Lyndon Baines Johnson and republican Barry Goldwater. Goldwater was a "hawk," with a reputation for supporting military aggression to ensure American dominance in the world. The Johnson campaign decided to exploit this reputation with the power of pictures, airing what turned out to be one of the most famous political ads in history.

A little girl is shown in a field of flowers, pulling off daisy petals, counting them one by one, birds singing in the background. When she gets to "nine," she suddenly looks up. A male voice says "ten," then starts a countdown sequence. When the voice reaches "zero," the scene dissolves into a horrific nuclear explosion. In a voice-over, Johnson says, "These are the stakes, to make a world in which all of God's children can live, or go in the dark. We must either love each other, or die." The tagline appears: "Vote for President Johnson on November 3rd. The stakes are too high for you to stay home."

In this campaign ad, a powerful, meaningful visual was presented to viewers before anyone could tell what the ad spot was even about—a classic example of picture before text, a classic example of PSE.

How can you apply the findings of PSE to your presentations? Use pictures in your slides whenever you can. Again, a picture is better at creating memories than text, as well as better at stabilizing them. It's a more efficient means of transferring information—worth a thousand words indeed.

But not just any old picture will do. We now know certain characteristics of images that make them even easier to pay attention to and remember. Here are two:

1. *Make the picture (or object) move.*

We lavish lots of attention on objects in motion. Bonus points if the object unexpectedly changes direction.

2. *Alter the characteristics of the picture.*

We pay tons of attention to objects that suddenly change color, alter in brightness (formally called *luminance*, the intensity of light bouncing off an object), or even if an object suddenly appears in our visual field.

Why put such premiums on changes to picture properties? As usual, there are evolutionary explanations. Consider motion: many of the experiences important to us in the Serengeti involved movement. Is there a rustle in the grass? Could be an ambush predator. Is there a sudden splash? Could be a tasty fish. Our brains are finely attuned to detect changes in the two issues very dear to our hearts: survival and food.

This is yet more evidence that we have hauled our evolutionary tendencies directly into the twenty-first century. We plopped them right into the middle of our PowerPoints.

The wandering eye

The American military is famous for making PowerPoint slides that are so complex, with hundreds of lines connecting to dozens of terms, that they resemble the root system underneath a tree. Almost always there's too much text, too much going on visually, too much of everything. As an example, there's a legendary slide labeled—get ready for a healthy dose of too-much-text here— "The Integrated Defense Acquisition, Technology, and Logistics Life Cycle Management System." It's composed of dozens of boxes filled with infinitesimal scrawls of English, too tiny to read, too much information to comprehend.

Such complexity is absurd to the point of being funny. When a similarly complex slide depicting combat dynamics in Afghanistan was shown to war commander General Stanley McChrystal, he quipped, "When we understand that slide, we'll have won the war."

That was in 2009. Twelve years later, American forces withdrew.

If you think too much text representing too much information is simply too much for human brains to absorb, you're in line with the peer review. Most experiments investigating PSE have involved comparing picture-retrieval with a similar stimulus presented in written form. Always, the verbal information is at the losing end.

Why is text so hard to comprehend? Research demonstrates there are at least two general reasons. The first involves busting a myth.

A lot of people think you read the same way you type, one letter at a time, one word at a time, marching lockstep in a consecutive sequence. Scientists used to think so too. This linear idea was even given its own name, the *serial recognition model* of reading.

The idea didn't have much of a shelf life. When reliable eye-tracking tech became available, researchers found that the eye acts like a soldier. A drunk one. It will start a sentence at the first word, then suddenly stagger to the middle of the sentence, pausing briefly to read the words. The eye might then skid backwards pausing, looking at words near the beginning again, then jumping, if it feels like it, to the end of the sentence. See why I say "drunk"? We call the forward jumps *saccades*, the backward jumps *regressive saccades*, and the pauses *fixation points*. The only reason you can read anything at all is that the *net* activity moves in the direction the writer intends.

It never ceases to amaze me that while reading my sentence, your eye is doing this very thing. Oddly, only parts of the staggering are mission critical. Word recognition primarily occurs at fixation points, when the eye is settled, which means the perceptual business of text only occurs when the jumping stops. This leads to a disturbing conclusion: while your eye is jumping, you are functionally blind.

Letter mess

You might expect that all this activity would make reading exhausting, and it does. But things are about to get much worse, mostly because of a minor contradiction that is still only partially resolved, a dispute that should convince you to always minimize the amount of text on a slide. It has to do with the limits of your highly adaptable brain.

We see many of the same words every day, year in and year out. Just think of the number of times you've read the word *the* today alone. You'd think that our highly adaptable brain would allow us to stop looking at the individual letters in familiar words, recognize that we've seen them before, then move on. Only when a word is unfamiliar should we stop to inspect its components.

That's exactly what does *not* happen. Your brain still must individually inspect each letter in each word, regardless of familiarity. Says researcher Deborah Moore:

> You might think that years of reading books, posters,
> computer screens, and cornflake packets would train
> the human visual system to recognize common words
> without the intermediate step of identifying letters ...
> Not so. A word is unreadable unless its letters are sepa-
> rately identifiable, and our reading efficiency is limited
> by the bottleneck of having to rigorously and inde-
> pendently detect simple features.

That doesn't mean familiarity isn't important, however, and therein lies our minor contradiction. It centers on your ability to read the following sentence, even though the letters are mixed up: The oredr of inidvidual leettrs in a wrod deosn't raelly mttaer.

The fact that this sentence is intelligible has been interpreted as evidence that brains really do recognize whole words, rather than individual letters. Does this fly in the face of Moore's findings?

Possibly. However, it's likely that while our brains are busy inspecting individual letters, they're also making comparisons with similar, previously encountered words. The brain can read the jumbled sentence fairly easily, but only as long as the first and last letters are correctly positioned (bonus points if there are strong context clues). This may mean there is letter inspection, familiarity inspection, and context analysis going on all at once.

Regardless of how this minor dispute is ultimately settled, the conclusion is stark: it takes a lot of effort to read text strings. Many processes have to engage simultaneously when we read words. And we have to try to make sense of them, using eyes that are roving over them like drunken sailors.

What to do with your letters

So, what does this mean practically? Volumes have been written about how to make effective PowerPoint slides. Some suggestions, though sourced from experience rather than laboratories, are quite insightful. Listed here are five things I think are worth noting. Though this quintet takes some of its advice from anecdote, it isn't five-alarm mythology. Indeed, most suggestions attempt to minimize the effort it takes to read text strings, probably because so much energy is wasted doing that deed.

Suggestion 1

Keep the size of a given typeface to 24 pt.

Suggestion 2

Limit the amount of text on a slide. Some professionals say 30 words, tops, spanning no more than 6–8 lines.

Suggestion 3

Watch your line length. A text string should have about 78 characters. This is the optimal size necessary to keep the eye fastened to its so-called "scan pat," a cognitive trail formed to keep the mind from losing its place.

Suggestion 4

Use a mixture of lowercase and uppercase letters, which means if you have a text stream, keep the caps-lock button off. Lowercase text is read faster than uppercase text, with speeds consistently 5–10% greater with mixed cases than with all caps.

Suggestion 5

Use sans serif typefaces, as opposed to serif, on all slides. As you may remember from an old design class, serifs are like little decorations on a given typeface, small flourishes such as lines or curls that dangle from individual letters. Serif typefaces are almost always more complex visually. (Times New Roman is a good example of a serif typeface.) Sans serif typefaces are those without those decorative serifs. (Helvetica is a good example of a sans serif typeface.) Eye-tracking experiments show that people read sans serif typefaces faster and with greater accuracy than their serifed cousins. With sans serif, people also experience fewer regressive saccades, those backward jumps the eye insists on doing. Interestingly, while this advice holds true for short strings of text, it does not hold true for paragraphs. Large blocks of text are better processed in serif typefaces than sans, at least on paper. But since large blocks of texts should never be seen on a slide, and slides are never composed of paper, this recommendation is added for completion, not for relevance.

There will probably be many more suggestions to make as this research matures. But for now, it's time for us to summarize what to do next Monday.

What to do next Monday

Suppose you have to give a sixty-minute presentation. What do the brain and behavioral sciences say you should do?

They say you should throw out the notion of a sixty-minute presentation. Your thoughts should be constructed instead around a half-dozen ten-minute presentations. You should have something cogent and memorized at the very beginning, given the rapidity and stability of first impressions. Preferably, it should be something introducing both you and your topic.

Secondly, research says to apply a hook at the ten-minute joints. The hook should be emotionally competent, relevant, and short. Bonus points if you can transform the hook into a narrative.

Thirdly, research says to pay attention to the visual aspects of your presentation, which for most people will be slides. There should be far more pictures than strings of text in each slide. Even more bonus points if you can animate them.

But hold on. I've ignored a certain aspect up until now that could potentially change the way presentations are conducted going forward. Almost every experiment described in this chapter was done in a pre-pandemic, let's-not-always-rely-on-Zoom-for-our-presentations workplace.

Many of you have had to give presentations virtually as a result of the pandemic. For some, it's been a temporary change. But if you find that you are giving presentations via computer permanently, does science tell you to alter any of the suggestions given here?

The answer is that nobody knows. Even given the broad impact of COVID-19 on remote interactions, distance-learning research is still remarkably thin, especially when compared with in-person presentations. All we're left with are personal observations. I'll close this chapter with some of my own.

Since the spring of 2020, I've done dozens of lectures and presentations remotely, and I've made some changes as a result. I've found that hooks are better placed every five to seven minutes during remote presentations. I use more slides too, or at least more animated builds per slide. Those builds always include objects in motion, with visual changes occurring every ten seconds or so.

I haven't changed everything, however, because some things *never* change. The brain still likes pictures, still craves narratives, still thinks movements imply either food or fear. After all, you can't beat back millions of years of evolution in the course of a single pandemic.

We started our discussion in this chapter with a commercial for a soft drink. It was no accident that it held your attention. It still can after all these years, even if you have to endure some cheesy '70s disco to feel it.

PRESENTATIONS

Brain Rule: *Capture your audience's emotion and you will have their attention (at least for ten minutes).*

- When it comes to your presentations, first impressions are important. Have the first lines of your presentations memorized.
- You have ten minutes to engage an audience. After ten minutes, if they aren't engaged, it will be much more difficult to get them to pay attention. By minute thirteen, if you haven't engaged them, you won't get them back.
- Emotions prioritize which inputs the brain processes. Stimuli attached with emotional appeals help audiences pay attention to your presentation and retain more information. The most powerful appeals contain emotions relating to threat and survival, sex (preferably the results of sex, like children), and humor.
- Sprinkle emotionally competent stimuli (ECS) or "hooks" into your presentation every ten minutes or so. These hooks need to tap the audience's emotions, be brief and relevant to the topic, and have a narrative structure.
- The brain is stimulated more by images than text or audible facts. Add pictures or even short video clips to your presentation when possible.

conflict/bias

Brain Rule:
Conflicts can be resolved by changing your thought life. It helps to have a pencil.

PUBLIC TELEVISION WAS GASPING for its last breath in the spring of 1969. Senator John Pastore, chairman of the US Senate Subcommittee on Communications, wasn't convinced of public television's worth. He was gripping the future of the public broadcasting world by its budgetary jugular, about to ax its $20 million subsidy.

The broadcasting officials of the time were understandably panicked. Happily, they had a secret weapon at their disposal. It was the humble form of Fred Rogers, legendary host of the children's television program *Mister Rogers' Neighborhood.* They invited Fred to testify before the subcommittee, hoping his words might save public TV. What occurred next is a master class in conflict management.

Fred started by describing what it was like to be a kid, and how his program related. "We deal with such things as the inner drama of childhood," he testified. He explained that he spent hours teaching little ones how to constructively manage their hot emotions, such as navigating sibling conflicts and the anger that arises in simple family situations. He pointed out that *Mister Rogers' Neighborhood* was valuable because it made clear that feelings are both "mentionable and manageable." And depicted people working out their feelings— producing a much more dramatic effect than usual TV depictions of adult conflict, which typically involved fists and firearms.

The magic in the room became palpable. His gentle kindness and emotional steadiness spread across the assembly as if Rogers were their pastor, and the Senate chambers his church. He invited the senators to experience the program through his testimony, asking them to be a "third person" by listening to a sample of its content. He read the lyrics to a self-authored song entitled "What Do You Do with the Mad that You Feel?" Rogers taught the whole room about executive function that day, especially impulse control, though he never used such words.

The strategy was effective beyond the broadcasting officials' wildest dreams. As Rogers finished reading the lyrics, Pastore melted like a warming glacier. "I think it's wonderful," the senator said, after confessing to goose bumps, eyes brimming with tears. "Looks like you've just earned your 20 million dollars." There was laughter in the room, followed by applause.

Rogers's interactions illustrate several conflict-management principles found in the behavioral sciences, though he may not have known it. We're going to discuss some of those principles, as well as the neurological mechanisms behind them. But we're also going to go further than just surface interpersonal disputes. In the second half of the chapter, we'll explore the deeper prejudices and biases behind many human conflicts, ones felt in workforces throughout the world. These biases are so powerful that, left unaddressed, they can last for generations and determine the fates of millions of people.

It's a heavy lift, for sure. We'll start simply, beginning with a few definitions.

Defining conflict

How does one characterize *conflict* in a meaningfully testable way? There are many forms of conflict from which to choose. Writers of fiction recognize seven different types of conflict; psychologists recognize four. There are internal conflicts (should I eat this pizza?) and external conflicts (should I start this war?). There are conflicts between intimate partners, and there are conflicts between complete strangers. For our purposes, we'll stick to conflicts dealing with workplace issues. (I'll leave dietary temptations, armed conflicts, and spousal and bar fights for another book.)

Social psychology defines conflict as "... the perceived incompatibilities by parties of the views, wishes, and desires that each holds." In the workplace, this conflict becomes interpersonal; it often occurs between people who (a) need each other to complete some common project and (b) hate that fact.

Harvard Law School defines three types of workplace hostilities between employees, each containing elements of the social psychologists' definition.

The first type are called "task conflict." These conflicts emerge when employees can't agree on how a job will get done (methodology), who will most effectively do the job (task allocation), or how much of the company's staff and assets will be earmarked for the job (resource dedication). These are relatively simple conflicts to address, because the issues are usually concrete and clearly perceived. Of course, *simple* isn't the same thing as *easy*.

Harvard Law calls the second type "relationship conflict." Relationship conflict surfaces when employees clash because of differences in ways of thinking, working styles, personality quirks, or even aesthetic perspectives. Since employees often have little say in choice of workmates, learned helplessness may rear its ugly head and be made worse because a paycheck's involved.

The third type is called "value conflict." This form involves clashes in ethics, morals, and deeply felt beliefs. Even employees' lifestyles can be brought into the conflict. Values are powerful forces in most people, an intrinsic part of their identities. Indeed, values are often found in a person's religious preferences or political points of view. Clashes in this category can, thus, be brutal—open hostility is common—for there is often no way not to take differences personally. It can even be the headwaters of toxic bias, a subject we'll visit in a few paragraphs.

The brains behind conflicting emotions

Task, relational, and value conflicts represent different experiences, but they share common neurological features. When the brain perceives a relational shootout is imminent, its survival circuits are marshaled to the front, regardless of conflict type. These circuits eventually generate emotions, ones powerful enough to affect productivity. Here's how Cor Boonstra, former board president of Philips, describes the role emotions play in conflict:

> *A company is a concentration of human emotions.*
> *Sometimes you want to achieve something, and you*
> *observe that it will not happen. Then, waves of emotions*
> *play a role, and bridges cannot be built. Most conflicts*
> *in organizations occur because emotions*
> *are not controlled.*

What types of emotions "can't be controlled?" What brain regions are involved in the affective experience of conflict? Boonstra is talking about the negative emotions—a rogue's gallery of discontentment, distrust, anger, and fear. Such feelings are by-products of the brain's activation of "survival circuits." They're fairly easy to visualize in the lab, mostly because the brain calls all hands on deck when it comes to peril perception. Indeed, we place our weightiest priorities on threats to our survival. Two pathways are simultaneously put on high alert, one called the *fast pathway*, the other the *slow pathway*.

The fast pathway involves the amygdala, a smallish, almond-shaped area deep in the middle of the brain. One of its many functions concerns something called *appraisal*. This is an evaluative process where the amygdala quickly decides if there's *really* something to worry about. If so, the little nut commands the brain to immediately issue threat alerts, activating those survival circuits we just mentioned. The activation occurs at such lightning-fast speeds, you're not consciously aware you're responding, hence the term fast pathway. This means your threat reactions are out of your control, at least initially.

They don't remain so, fortunately, and that's because of that simultaneously stimulated slow pathway. When activated, these circuits—which are found in cortical structures immediately behind your forehead—function as an inspector general over the amygdala, determining if the threat merits a larger response. If your cortex agrees with the amygdala's threat assessment, the cortex will command the amygdala to stay on post, maintaining an alert status felt throughout the body. This evaluation takes time—cortical regions have vast thickets of neural underbrush—which is why it's called the slow pathway.

Your brain's angels and devils

Fast and slow pathways aren't the only ways your brain responds to conflict. When you're prepping for threats, you're also making a social evaluation about your opponent. And your brain responds in a way you might find embarrassing: as if it were part of a TV cartoon trope. You've probably seen this trope many times, especially in old *Looney Tunes* episodes. The protagonist needs to make a moral decision, so an angel appears on one shoulder, a devil on the other. The angel tries to convince the protagonist to do the right thing; the devil advocates yielding to the dark side.

Your brain recruits its own versions of these cartoon specters in conflict situations, reacting beyond the initial peril activations. Let's suppose you're in an argument with someone you perceive as an enemy. Your shoulder devil immediately takes center stage, deactivating an area of the brain called the *anterior insula*. That's a big deal and also a bit strange. The insula normally receives physical information about where you are in relationship to your world (just like an internal GPS) and how you currently feel about it (like an internal psychiatrist). Taking them offline hobbles their function, of course, though exactly why is a bit of a mystery.

The devil isn't done with its work. As the insula begins to power down, two other regions power up, called the *nucleus accumbens* and the *ventral striatum*. The nucleus accumbens is one of the regions involved in metabolism of dopamine, that neurotransmitter involved in mediating pleasure and reward. The ventral striatum, which uses dopamine too, is involved in decision-making. Quite literally, the suffering of someone with whom you are in conflict doesn't always elicit horror in you. Sometimes it elicits a smile.

Like I said, a shoulder devil.

By the way, this does not occur if the person in distress is a friend. In that case, the shoulder angel takes center stage. Instead of the insula powering down, it actually powers up, allowing you to know exactly where you are and how you feel about it. Conflict, it turns out, is not a simple experience to the human brain.

Ballooning with empathy and compassion

For years, I consulted with the Boeing Company, often guest lecturing at their Boeing Leadership Center. I had a wonderful time discussing the interface between design engineering, human factors, and the cognitive neurosciences.

A conflict I commonly experienced at Boeing was the near-constant tension between research and development engineering groups and the people who managed them, a common occurrence in product-driven tech companies. Both sides would often joke about it, one variation being particularly memorable. The joke involved a person in a hot air balloon, 30 feet in the air, who shouts to an engineer on the ground below him.

"Excuse me. Can you tell me where I am?"

"Yes, you're in a hot air balloon about 30 feet above ground," the engineer responds.

"You must be an engineer," the ballooner says.

"I am," the man says. "How did you know?"

"Everything you told me is technically correct, but it's of no use to anyone," the ballooner snarkily replies.

The engineer also becomes indignant and says, "You must be in management, sir."

"I am indeed. How did you know?"

"Easy," the engineer responds. "You don't know where you are or where you're going, but you expect me to be able to help. You're in the same position you were in before we met, but now it's my fault!"

Though the joke is funny, the consequences of conflicts are not, and the impact, which can quickly escalate, can begin to hurt productivity. What can be done to reduce such tension? Interventions would be nice, obviously, and the business shelves in bookstores offer plenty of self-help titles from which to choose. Could any of them speak to a Boeing engineer? In other words, do any of them work?

The behavioral research world has shown that certain interventions can indeed be quite effective, sometimes dramatically so. We're going to examine some of these interventions, which include exercises in perspective-changing as well as physical actions (one literally involving a pen and paper). The steps are sourced from twin foundational behaviors we'll need to define scientifically: empathy and compassion.

Empathy and compassion feel like similar experiences, so you might be surprised to know the behaviorists make a distinction between them. Empathy involves a subjective experience, which we've previously discussed. Compassion involves a desire to help, usually resulting in an action, which we haven't discussed at all. Compassion is where all the practical solutions lie.

Here's a way to distinguish between them. Suppose on a business trip you lost your hotel room key. You consult the front desk, explaining your plight. Further suppose the clerk responds only with empathy. He might say, "I'm so sorry. It's awful when you are in a strange city and can't get to your room," and then turns to help another customer. While this empathy's a nice sentiment, it doesn't help your dilemma, especially if that's all he communicated. If the clerk responds with compassion, however, he might still empathize but would also feel a desire to help, which usually translates into an action. "No problem," he might say. "Let me make you another key." He then hands you a new card, solving your problem.

That compassionate *bonhomme* has a helping component. That's how you tell the difference. Empathy triggers an urge to *feel*, while compassion triggers an urge to *do*. The steps we are about to discuss are filled with actionable conflict-managing ideas. The research world calls them *deeds of compassion*, simply because they are designed to help in an urge-to-do fashion.

Defusing conflict

We now turn to three evidence-based urge-to-do intervention protocols. They all involve taking an action to deal with emotions. As Boonstra, former chair of Philips, suggested, conflict usually produces negative emotional responses. These often have a silver lining, however. Sometimes they're the bridge to making things better, the beginning of a compassionate response. That's key. If you don't deal with emotions, you won't get anyone to sign a peace treaty. According to one investigator, "… emotions are critical elements of conflict … emotions mediate the relationship between perceptions or appraisals of conflict and conflict resolution strategies."

The first evidence-based protocol involves your imagination, specifically your ability to *predict* the emotions of others. The second involves your ability to *detect* the emotions of others, to see if your forecasts are correct. The third involves your ability to *control* emotions—not the emotions of others but of yourself—once you see more clearly the emotional terrain.

Let's take these one at a time, beginning with the first group of protocols. Scientists call them "experience-based." These entreat you to use your imagination to speculate about—and thus understand—the internal emotional lives of your opponent's world. For example, the perception can originate from pure mental transposition. ("Consider what it feels to take a walk in their shoes.") It can also be based in past experiences with your opponent, sourced from memories of your opponent's behavior, suggesting a map of their psychological interiors.

216

The second group of protocols is called "expression-based." These involve your ability to recognize the actual emotions of your opponent as they are happening, rather than simply trying to theoretically imagine them.

In one remarkable pre-post experiment, resident physicians watched a series of empathy-focused videos developed by behavioral researchers. These videos (a) taught residents how to detect subtle changes in facial expressions in their patients, (b) instructed them on the neurobiology of empathy, and (c) showed them film of more-experienced doctors interacting with their patients. Both the experienced doctors and their patients were rigged up to equipment measuring their real-time physiological responses. The results were displayed on a corner of the screen, so the residents could see for themselves what empathic or nonempathic responses actually did to a patient's body.

The results were extraordinary. The residents' ability to recognize emotional cues—subtle or not, verbal or not—got dramatically better, as did their empathic understanding. Most critically, their patients' satisfaction with the residents' demeanor improved, measured by something called a CARE test. (This rating functions as a report card, with patients evaluating both empathic and relational skills of their doctors.) Improvements were still evident two months after the training ended.

Both experience-based and expression-based protocols ask you to ignore your own psychological interiors for a period of time. The third and final intervention asks you to do just the opposite.

The world of Pennebaker

This somewhat inward-looking conflict-resolution protocol involves an external action on your part, the kind which compassion would motivate you to commit. As you recall, a desire to help is one of the signatures of compassion. It leads to an external action step. So, what compassion-powered action steps do you need?

For answers, we turn to social psychologist James W. Pennebaker, a researcher who loves words like some people love milkshakes. His talented family loves words too. His wife, Ruth Pennebaker, is a famed columnist and author. His daughter works in communications in Washington, DC. Though Pennebaker is an internationally recognized behaviorist, author of more than 300 peer-reviewed papers, he still writes books with titles like *The Secret Life of Pronouns*.

We're fortunate his lexical love affair continues. Pennebaker is one of those rare behaviorists capable of transforming ivory tower research results into main street practical therapy. He's probably best known for something called *expressive writing*, one of the most potent interpersonal conflict-management tools on the market. And, of course, he's written many words about it.

Pennebaker was the first to discover that writing about experiences that bug you causes them to not bug you after a while. But you have to write in a particular way and with a particular periodicity. His work was originally designed for people who have experienced trauma. It was eventually found to remedy people with many other kinds of tough emotional experiences besides trauma, including people in deep need of conflict resolution. It is the third of our three conflict-resolving protocols.

To understand Pennebaker's expressive writing protocol, which concerns writing while shifting points of view, we need to talk about the difference between self-immersed and self-distanced perspectives.

One of the most common post-fight experiences involves a rumination phase. It's usually a fantasy, where you reimagine the conflict, perhaps mentally rewriting the experience to make you look better (such as biting, rapier-like *I-should-have-said-this* retorts, for example). This rumination is invariably done in the first person, through your own eyes, a viewpoint researchers call the "self-immersed perspective." Research shows that such immersions increase negative feelings and resolve virtually nothing.

Fortunately, behaviorists such as Ethan Kross have found you have another perspective choice in your rumination phase. Suppose that, when a conflict occurs and the rumination editors in your head start rewriting the conflict from your point of view, you fired them? What if you replaced them with videographers recalling the conflict from the perspective of a neutral third party filming the event? This approach necessarily pushes you out of the role of intimate participant and into the role of dispassionate observer. You're a person with a camera, not a wound. This shift is called "self-distanced perspective."

It's a familiar idea. Self-distancing is the secret sauce behind the advice most marital therapists give their clients to reduce conflict. They instruct couples to keep away from *you* statements—such as "You forgot to lock the door"—which tend to be accusatory. Couples are instructed to reframe statements by appealing to more neutral verbals: "The door was unlocked, and I was afraid someone would steal something." This reframing is the voice of a videographer, not of a protagonist. Such reframing has been shown to decrease conflict. Indeed, it is part of a parcel of ideas behind Pennebaker's *The Secret Life of Pronouns* book.

Self-distancing doesn't come naturally to most people, so most people don't do it. The combined insights of researchers like Kross and Pennebaker show us we probably should, and from a hybrid of their ideas, we can detail our action steps.

The pen is mightier

For purposes of discussion, let's assume you've just had an interpersonal conflict at work, an intense verbal altercation with a colleague that's left you fuming. You know you're going to have to work it out. Let's apply the expressive writing mash-up we just discussed—blending the ideas from both Kross and Pennebaker—to see if you can increase the probability of resolution. The protocol can be described in three steps:

1. *Timing*

Carve out twenty minutes per day for the next four days.

2. *Writing*

During those twenty minutes, write about what happened. Do so from the perspective of a third person observing the conflict—the videographer's perspective. Write what your "opponent" said or did. Write what you said or did. Include descriptions of your intentions and motivations during the conflict, as well as those of your opponent. If it helps, observe the old adage "people do what make sense to them." Write about what made sense to your opponent at the time, and then write about what made sense to you. This can be hard, at least at first.

3. *Repeat the exercise for the next three days, spending twenty minutes each time, rewriting the experience.*

For expressive writing to work, one needs to follow a few rules. First, write continuously, in stream-of-consciousness prose. Don't worry about spelling or grammar or punctuation or any other "bone" getting in your way—free-range words are the tastiest.

Secondly, write for an audience of one: you. Your writings should not be part of a tactical offensive strategy to crush your opponent, nor an effort to convince some imaginary jury why you were right and your opponent was wrong. Remember, you're a third-party videographer describing things from a detached point of view. Some people tear up their writing when they're done. Others keep their prose to review later. Whether the writing is destroyed or archived, the intent is for only one person to ever read it.

These techniques allow you to distance yourself from what happened, preparing you in advance psychologically for the time when/if you must actually confront your opponent directly. The benefits are legion. Psychological well-being, overall health, and physiological functions—especially those related to stress—demonstrably improve with this technique. Positive resolving results are still evident three months after the writing stops. Meta-analyses, a type of paper that asks "Is this consistently true?" over a broad swath of research, shows the positivity in dramatic fashion, sporting R-values in the range of 0.611 to 0.681. (R-values, as you might recall from an old statistics course, measure correlative relationships between two variables, the higher the R-value the better—and 0.6 is a solid mark.) Pennebaker's exercise breaks the fever that causes so many relationships to become sickly.

That's the secret medicine. Remember, negative emotional responses are the major impediment to resolving most conflicts. Reducing the emotional temperature is, thus, a critical step. Keeping the emotional temperature low is even more valuable, especially if your goal is long-term resolution.

Conflicts and biases

I promised earlier to take a deeper dive into the nature of conflict, discussing the prejudices, stereotypes, and biases that historically have fueled some of our most appalling behaviors. I'd like to begin with Shakespeare, quoting what, to me, is the most wrenching, ghastliest speech the bard ever penned. The words were uttered by Shylock, one of Shakespeare's most controversial characters, from *The Merchant of Venice*:

> *I am a Jew. Hath not a Jew eyes? Hath not a Jew hands,*
> *organs, dimensions, senses, affections, passions? ... If*
> *you prick us, do we not bleed? If you tickle us, do we not*
> *laugh? If you poison us, do we not die? And if you wrong*
> *us, shall we not revenge?*

I say "wrenching" because the prejudices Shylock describes are so cruelly unfair, and I say "ghastly" because the anguish he utters is still so horrifyingly relevant. One need look no further than the Holocaust to understand the hideous power bigotry can hold over human beings. Or to the Rwandan genocide in Africa, or to our own shores, where 400+ years of slavery, Jim Crow, and racial inequality created an open wound that has yet to close. Bias is a sticky, pernicious substance. It oozes into almost every corner of life, from religion to ethnicity, from politics to gender, from the way we think about old people to the way we think about fat people. As you'll see, frustratingly, bias is extremely difficult to measure. You'll also see it's extremely difficult to wash off.

Definitions

From where do such sticky behaviors arise? Nobody I know likes them. Nobody I know wants them. Yet everybody I know has them. Research shows everybody you know has them too. We won't have much purchase on the answer until we define a few terms.

Scientists file biased behaviors under the obscure category of *social motivations*. These are the forces compelling us to organize our affiliations with others (the ability to distinguish "me" or "us" from "them" or "others," for example). This organizing need may initially have been a selected-for trait, evolutionary psychologists declare. It fueled our desire to create social groupings, then characterize them. The tendency created alliances capable of conquering the world, then conquering each other.

Things become dangerous when people ascribe value judgments to the groups in which they claim membership. "Us" (sometimes called the *in-group*) develops into "safe" and "terrific," whereas "others" (sometimes termed the *out-group*) become "unsafe" and "less terrific." Team loyalty isn't far behind, and tribalism—defined by social fealty—becomes as common as soccer fights in Europe.

This social rubric allows us to understand the scientific definitions of three important concepts, ones that also carry nonscientific, layperson definitions:

Stereotype

Stereotyping involves the word *overgeneralization*. People who form stereotypes of an out-group look for characteristics (behavioral, physical, economic—the list is painfully long) that appear repeatedly in that out-group and ascribe them to every member of said out-group. Ethnic jokes often use these characteristics in their attempts to land punchlines.

Prejudice

Prejudice involves any emotional reactions—termed *affective responses*—that members of an in-group might carry toward members of an out-group. The reaction could be hideously affirming (all blond-haired people are wonderful) or terrifyingly negative (I hate all Jews). Stereotypes are formed in the mind. Prejudices are formed in the heart.

Bias

Biases are differentiated from the other concepts by their involvement with perceived threats. Biases come in two flavors. The first is called *explicit bias*, which is commonly defined this way:

> ... the attitudes and beliefs we have about a person or group on a conscious level. Much of the time, these biases and their expression arise as the direct result of a perceived threat.

The second flavor is called *implicit bias*, a judgment that flies just below the conscious-awareness radar. Even blind to its presence, its owner still reacts to its "threats." Also called *unconscious bias*, implicit bias is defined as "... the brain's automatic association of stereotypes or attitudes about particular groups, often without our conscious awareness."

For many, the presence of an implicit bias is an embarrassing, invisible scarlet letter, a trait at odds with their own value systems. Yet these biases seem to exist, lurking just below our awareness, swimming in a psychological ocean that used to be called "the subconscious."

The IAT

If implicit biases exist mostly out of our awareness, how do we know they exist? Is it like an odor that nobody can smell? Behaviorists have ways of measuring implicit biases, however subtle they may be. One famous—and controversial—psychometric instrument is called the Implicit Association Test, or IAT.

I will tell you flatly that some people are furious with this test, especially after they take it. Reactions range from feeling falsely accused to feeling acid-hot humiliation. If the test reveals something negative about you, and you can't see it, you may find yourself disagreeing with it, throwing stones at it, and ultimately, disbelieving it.

Let's find out what all the fuss is about.

The IAT is a multipart exam that measures how strongly you feel an association exists between two variables. The first variable is called a target (such as young, old, or middle-aged people). The second is a called an assessment (good, bad, indifferent). Executed by a computer, the IAT flashes certain word pairs, then measures the time it takes you to agree with a specific association. Since we have shorter reaction times when we feel certain that concepts fit with each other, measuring that timing theoretically reveals something about the way we think.

Though the test is actually more complex, here's an illustrative example. Let's say you struggle with ageism, that you have an implicit bias toward young people and against the elderly. You would respond more quickly to a pairing of the words *young + good* or *old + bad* when these flashed on the screen than you would to a pairing of the words *young + bad* or *old + good*. Though you may consciously feel everyone is equal, the test indicates an implicit preference for younger citizens—an implicit *bias*.

Note that an IAT can assess many things besides ageism. It can be used to assess attitudes toward race, sexuality, gender, and religion, to name just a few.

You should know there have been scientific objections to the IAT. Researchers have raised concerns about its test-retest reliability (the ability to obtain a score at one time point, then obtain a similar score at another time point). There have also been criticisms about the undue influence certain social contexts exert on IAT scores. And, as ever, there are vulnerabilities involving people trying to game the system—usually taking the form of respondents trying to answer questions according to what they think the researchers prefer (or what they themselves want to believe) rather than according to what's actually resident.

Most of these objections have been addressed in one form or another, and the current consensus is that we should not throw out the baby with the bathwater. For example, the IAT is extremely good at predicting how people will vote. The cautious endorsement of the research community may be summarized in the title of a recent paper investigating the IAT's strengths and weaknesses: "The IAT Is Dead, Long Live the IAT."

I couldn't agree more. I'm happy to continue describing research using the IAT, as long as we keep the aforementioned issues in mind.

One important finding the IAT uncovered was just how hard it is to change a bias once it has burrowed its way into our psyche. Muhammad Ali once waxed eloquently on this persistence. The great boxer was interviewed in front of a live audience by BBC journalist Michael Parkinson. Ali talked about his childhood and his dawning realization of racial bias:

> I'd always ask my mother. I'd say, "Mother, how come is
> everything is white? ... I was always curious. I always
> wondered why, you know, Tarzan is the king of the jungle
> in Africa. He was white.

The audience broke into sporadic, nervous laughter.

> *Everything was white. And the angel food cake was the*
> *white cake. The devil food cake was the chocolate cake*
> *... everything bad was black. The little ugly duckling was*
> *the black duck. And the black cat was the bad luck.*
> *And if I threaten you, I'm gonna blackmail you.*
> *I said "Mama, why don't they call it 'whitemail?"*
> *They lie too. I was always curious. And that's when*
> *I knew something was wrong.*

Indeed. It's been more than half a century since Ali did that interview, and we still deal with these issues. Biases have a way of wriggling into our thought life, then adhering to us like parasites.

Role of awareness

Many people recognize there's a problem these days, fortunately. And there have been attempts to institutionalize remedies. An entire cottage industry has sprung up around the issue, creating programs such as "diversity training," purchased by HR departments around the world.

Some trainings claim in their advertising to reduce or even eliminate implicit bias. You'd do well to be skeptical of such claims. When you look for supporting data, few pre-post studies detect reliable conversions, and the ones that show real change demonstrate laughably short durability. Changes in bias last less than an hour or, for some, less than half an hour. Most programs don't even have the luxury of being tested. This was pointed out by the researcher who invented the IAT, Anthony Greenwald, who said in a recent interview:

> *I'm at the moment very skeptical about most of what's*
> *offered under the label of implicit bias training, because*
> *the methods being used have not been tested scientifically*
> *to indicate that they are effective.*

This is heartbreaking. Many diversity training programs are designed by well-intentioned, good-hearted people who see open social wounds and want to heal them.

Happily, there are grounds for hope.

We're going to examine a number of approaches that have some claim on being promising. All have had the luxury of undergoing peer-review. Our first strategy is to find a way to ignore our biased tendencies, stepping over the mess on cognitive Aisle X and moving on. One remarkable study, now more than twenty years old, did just that. It examined, of all things, sexism in American symphony orchestras.

Lessons from the symphony

Symphonies used to fill positions by auditioning musicians from pools selected by the music directors, who historically were male. Not surprisingly, these music directors selected mostly male players. By the end of the 1960s, only 6% of musicians in the highest-rated symphonies were female.

Things changed as the years went by. Certain symphonies began holding their auditions with candidate musicians *behind* a screen or curtain. This blinded the sex of the candidate to the selection committee, which also blinded the committees' biases. The effect was profound. By 1993, the number of female musicians in the symphonies had climbed to 21%.

Such studies are often pointed to as bias-busting success stories, but there's a problem. Strategies like the blind auditions deaden the effects of biases without doing anything to actually get rid of them. They sever the stem of the weed but leave the root intact.

Is it possible to take a cognitive weeding shovel to our thinking and unearth the root of the bias itself? Happily, evidence of this possibility exists. I found a clear example in a BBC radio broadcast installment entitled *Two Minutes Past Nine*. Its subject was the horrific Oklahoma City bombing that occurred in April of 1995. The last episode of this series documenting a sad chapter of American history discussed the reconciliation work of Imad Enchassi, the imam of a mosque in Oklahoma City. The mosque—really a complex of facilities (including a free medical clinic)—was regularly picketed by anti-Islamic militias.

One day, the imam summoned enough courage to walk over to one of the protesting militia members, a strapping bull of a man brandishing an M16. The cleric asked him why he was protesting, expecting to hear the equivalent of an all-caps tweet-rage. The man replied simply that he was "demonstrating against Islam," which began an extended conversation.

The imam soon noticed the protester had a suspicious-looking mole on his face. He suggested the man get it examined. The protester related he had no money, and at that, the imam brightened. "We have a free clinic right here!"

He then led the militia man, still brandishing his M16, into the building. Sure enough, the mole was cancerous, in need of treatment. Something changed in the protester that had staying power. To make a long story short, the imam said, "… he's being helped to this day by our free clinic. And now, he is a guy that does security for us."

Can you train people to have behavioral epiphanies like this militia man? What kind of wisdom could bring to scale the kind of grace Enchassi extended to the protester? Is it possible to inject healing into the bloodstreams feeding our entrenched biases? A number of research efforts have endeavored to do just that.

Approaches being tested

Three evidence-based approaches have shown success and continue to be tested. Here they are:

1. The go-slow approach

These approaches train people to slow down before they make decisions that could be potentially influenced by bias. Researchers found that people who reacted quickly to a given situation usually reacted out of their internal preconceptions, but if allowed to think slowly before deciding—meditating, internally deliberating—they tended not to act in as biased a fashion.

2. *Environmental approaches*

These approaches are based on the finding that our biases aren't on all the time but can be triggered by specific environmental cues. If the triggering cues could be isolated, biases could be dragged out of the dark, the obvious first step toward getting rid of them. (It's that critical step of *ownership*.)

3. *Education*

Strategies have ranged from describing how bias works to providing counterprogramming examples of successful people who don't fit a given stereotype. Two collaborating research groups, seeking to counter sexism, noticed that teaching people about sex-based behavioral differences improved mentoring relationships between men and women, for example. Writing in the *Harvard Business Review*, they reported:

> *Men are more self-aware and effective in mentorships*
> *with women if they understand and accept ...*
> *the neuroscience of sex and gender, and the equally*
> *powerful impact of gender socialization.*

Sadly, we don't know if education is the right strategy for successfully treating every bias out there. We also don't know the long-term effects of go-slow approaches or social-triggering practices. That no magic bullet exists for every type of bias is hardly surprising. We have made remarkable progress in settling formerly hot-button legal issues like gay marriage, but we have yet to make as convincing progress on other socially toxic biases. (Race relations come to mind.)

Yet there's reason for hope. Most training strategies involve addressing, even remodeling, someone's thought life. That's where the cognitive neurosciences may be helpful. We actually know how to provide convincing alternatives to the way people think and feel about things. You won't find the knowledge in a training program, however. You'll find it in a therapist's office.

Change your mind

Occasionally you'll run into a titan of science who looks—and acts—nothing like his stature. Case in point is the legendary developer of *cognitive behavioral therapy* (CBT), psychiatrist Aaron Beck. He has a self-deprecating demeanor, sports a shock of white hair, and wears a bowtie right out of a 1950s comic book. His voice is thin as a reed, his mind as sharp as a scalpel, his heart as warm as a holiday fireplace.

Beck's CBT has probably relieved more mental misery than any other technique in the history of psychotherapy. And it's based on a simple insight: creating long-term behavioral change involves attacking the negative behavior at its source—a person's thought life. Based on this seemingly obvious idea, CBT results in quantifiable behavioral changes, sometimes measurable in years. Not surprisingly, CBT is used by clinicians all over the world.

I'd like to get into the crawl space underneath Beck's protocol, describing its foundations before returning to our discussion of bias. We'll start with a subject that CBT has proven particularly adept at addressing: anxiety.

Suppose you were afflicted with impostor syndrome, a type of anxiety many successful businesspeople experience. With this anxiety, you feel that (a) you are woefully incompetent at your job, (b) you got as far as you did by sheer blind luck, and (c) at any moment, someone's going to find out. CBT directs you to pay attention to your thought life; you're going to use that instruction to get rid of impostor syndrome.

This is what CBT says to do:

1. *Identify the NAT.*

NAT stands for *negative automatic thought.* Your first goal is to isolate and identify the source of the anxiety. In the case of impostor syndrome, the NAT is the feeling of being found a fraud.

2. *Evaluate the NAT.*

CBT asks you to assess the legitimacy of the NAT. "What makes you feel the NAT is true? What's the evidence?" CBT inquires. "Are there perhaps more balanced alternatives to the NAT?" In the case of imposter syndrome, the alternative could be: "Wait a minute. I'm not always a fraud. I actually do a few things well." CBT never asks you to believe the alternative, interestingly, only to conjure one up. In fact, belief isn't the issue. But habits are, as the third step explains.

3. *Reward the alternative.*

CBT then asks you to do two things. First, start a pairing exercise. Every time you think of your self-defeating NAT, allow its presence to trigger the less self-defeating alternative. Every time you think you're a con, bring up the idea that you're not a con. Second, every time you pair the two, give yourself a small reward. The reward could be anything. (One colleague I know popped a jellybean into her mouth every time she successfully paired the two.) The reward has to be truly pleasant and consistently supplied.

Research shows that, where CBT is applied dependably, NATs wither of their own accord in time, leaving you with only the positive alternative (and, in the case of my colleague, a few more calories). CBT has proven to have powerful effects in a wide variety of psychopathologies, ranging from depression to obsessive-compulsive disorder, even schizophrenia.

A promising direction

As our CBT discussion suggests, evidence-based techniques exist that are potent enough to change one's mind. Could one use the power of CBT to reduce racial implicit bias?

Researchers headed by Patricia Devine at the University of Wisconsin–Madison decided to try. They devised an intervention with a quintet of behavioral subcomponents, several of which involved CBT-like exercises.

One Beck-esque example was called *stereotype replacement*. The subjects first identified their target NAT, in this case the response to a self-generated racial stereotype. Subjects were then instructed to consider an unbiased response followed by replacement instructions, swapping out their muddy NAT for a cleaner, less biased model. This training included instructions about how to avoid future stereotype-based responses.

With the intervention design in place, the experiment could begin. After first obtaining IAT scores from all subjects, the subjects were randomly assigned to control and experimental groups. The experimental subjects underwent training sessions using the researcher's five-part intervention.

Devine then retested everybody's IATs at several points post-training, including spots at the fourth and eighth weeks. Remarkably, Devine found the needle was already moving in the experimental group by the fourth week. Though the entire project took twelve weeks to finish, by the eighth week, she had her answer. It's best summarized by a quote from the paper:

> ... *people who received the intervention showed*
> *dramatic reductions in implicit race bias. ... People*
> *in the control group showed none of the above effects.*
> *Our results raise the hope of reducing persistent and*
> *unintentional forms of discrimination that arise from*
> *implicit bias.*

That's quite a thing to publish—and quite a thing to contemplate. Biases might behave like they are made of quick-drying, long-lasting cement, but you can still make deep cracks in them. They are even amenable to long-term removal, at least if you use Devine's jackhammer, with positive effects still measurable months and years (as opposed to minutes and hours) after the training finished. The most robust outcomes were reserved for those subjects already concerned about racial bias before entering the program. Yet everybody in the experimental group experienced an increase in their general concern about bias. They also experienced an increased awareness of their own predilections toward stereotyping.

Since the original publication of Devine's work (2012), replications have reassuringly confirmed the initial claims of success. Two years after the intervention, experimental groups still showed measurably stronger anti-biasing instincts than controls, including the willingness to confront biases they encountered in social media. The intervention also affected gender balancing in the hiring practices of academic departments. In units that had undergone the training, 47% of the new hires were comprised of women. In controls without the training, that figure stalled out at 33%.

The program has been effective enough to have been codified into a workshop, sporting a name obviously not approved by the UW–Madison marketing department. It's called the "prejudice habit-breaking intervention" program.

And that brings us to my number one recommendation about what to do next Monday. Investigate this program. Learn the behavioral underpinnings of its evidence-based design and become an expert in CBT-based interventions. If helpful, use this chapter as a guide. A detailed description of this workshop can be found at breaktheprejudicehabit.com.

The reason is simple. Devine's work gets to the root of the bias problem: somebody's thought life. We have our work cut out for us, of course, but treating the condition at its source is something the behavioral sciences are really good at doing. That offers true hope. One day, protocols like Devine's may be capable of making Shylock's plaintive cry a museum piece of how we used to think, rather than an illustration of just how far we have to go.

CONFLICT/BIAS

Brain Rule: *Conflicts can be resolved by changing your thought life. It helps to have a pencil.*

- The first step to containing a conflict of any kind at work is to recognize your opponent's emotions and control your own emotions, especially discontentment, distrust, anger, and fear.

- If you get involved in a conflict, take twenty minutes a day during the few days afterward to write down what happened—but from the point of view of a neutral, third-party observer. This will help you to decenter from the situation.

- Remember that certain conflicts can stem from biases between two parties' different morals, deeply held beliefs, or lifestyles.

- Be wary of any HR training programs that claim to eliminate bias. Evolutionary psychologists believe biased-based behaviors were a selected-for trait, helping humans create social groupings for survival. As a result, biases are almost impossible to eliminate, both for individuals and groups.

- Though not perfect, the Implicit Association Test (IAT) is useful for uncovering implicit biases and predicting some behaviors, such as how people will vote.

- Consider the prejudice habit-breaking intervention program from the University of Wisconsin–Madison, which has shown the most promise in uprooting biases by using components of cognitive behavioral therapy to change thought habits.

work-life balance

Brain Rule:
You don't have a "work brain" and a "home brain." You have a single brain functioning in two places.

"Here's to the beginning!" Dolly Parton prophesizes, effervescent as a can of soda pop.

Those sparkling words are nearly the last lines spoken in the iconic 1980 movie *9 to 5*. Parton's character, Doralee Rhodes, has reasons to be happy. Wrenching control from the office's resident alpha male, she and two of her fellow female colleagues presided over an office takeover or, more accurately, office makeover, instituting changes that seem radical even today. These modifications included onsite day care, flexible working hours, job-sharing, a substance abuse rehab benefit, and a plea for equal pay. Such alterations cut down on office absenteeism, noted the script, and resulted in a 20% rise in productivity. Rhodes made her proclamation when the board chairman heard about the success and came down for a personal inspection.

The movie was a big hit, fueled by the equally famous Parton-written song with the same title. It also spawned a TV show that ran for five seasons, pinching a cultural nerve that still seems relevant today.

That pinch is the focus of this chapter: the balance between work life and home life. Balance may not be the best name, however. In the United States, it's not so much trying to find an equilibrium between two competing priorities as it is like trying to sign an armistice between two warring factions. The pandemic of 2020 has changed the parameters of work-life issues, but the battlefront still exists and the negotiations are still ongoing, even after all these years. Those suggestions from 1980, sadly, have still not come true.

There is hope, however. We'll explore who the combatants are and what brain science suggests the peace treaty should look like.

We'll discuss the hard part of the negotiations: managing stress, which is really about managing control. We'll continue with a discussion about how home life—from nurturing partnerships to nurturing kids—can be a great friend to business life. And we'll end by discussing what it's going to take to make Rhodes's heartfelt toast to the future not just a late twentieth-century aspiration but an early twenty-first century reality.

Not surprisingly, we have our work cut out for us.

Home/work

So, what do we mean by work-life issues, and why are they so hard to balance?

As mentioned, one reason is that they're a struggle between priorities, which mostly comes down to managing the clock. There are the demands of work on one hand, the demands of home on the other, and the same timer ticking against them both. When the professional and the personal don't get in the way of each other, life can be manageable.

In the real world, however, life is seldom manageable, at least in North America. Labor and life compete for employee attention and energy, sometimes fiercely so, especially if one adds raising a family as a priority in the life column. Many people stagger between the competing priorities as if fighting a hangover. For sleep-deprived workers with newborns, sadly, there is no way of hydrating one's way out of the pain the next day.

What's usually sacrificed in these balancing attempts is the psychological health of the employee. It's one of the reasons why cognitive neuroscientists are sometimes called in by their business school colleagues for insight. The starting place for both sets of professionals concerns how they define what they're talking about.

Business professionals use terms like "work-to-family inter-ference," for example, which is defined as something that happens at work which influences the family: Your boss yells at you at a meeting, so you go home and kick the dog. Or you get a promotion at work, so you take your family out to dinner. The term "family-to-work interference" defines its opposite: Your daughter spit up all over her bed, causing you to be late for work. Or your partner wrote you a love note, and then you breezed through your presentation.

Note that the above definition of family and workplace implies that everybody agrees on what those terms mean—not a good idea. It's been a long time since people could treat shows like *Leave it to Beaver* as documentaries. Half of all babies are born to unmarried couples, for example. Many kids are now raised in single-parent households. Nontraditional families include same-sex and gender-nonconforming couples. The list is broad, growing, welcomed, and to some people, uncomfortable.

The workplace is experiencing similar definitional shifts. These shifts started before the pandemic and greatly accelerated afterward. COVID-19 blew up the concept of the workplace residing in a far-from-home building. It is likely that many will continue to include "home" in the "office" experience.

The fact that these changes are so new means there's a scarcity of research investigating how nontraditional households and nontraditional workplaces function in society. The little that's out there shows remarkable consistency with older research studying "traditional" family and work structures. But until more time has passed, we will have to fasten this paragraph to science's favorite footnote: we need further research.

What you *can* say is that living on the work-life balance beam is changing. As is true with most social disruptions, these changes are sources of tension. When brain scientists work with business researchers on issues related to mental health, one of the first things they discuss with each other is how brains respond to stress. They don't get very far with the collaborations in this area until they define exactly what they mean by *stress*.

Stress at one level is easy for both sets of professionals to understand. It's the ingredient where workers who have tried to "do it all" find they can't even "do it a little."

Stress and its effects

So, let's talk about stress.

As you probably know intuitively—and researchers know quantitatively—stress can really hurt cognitive function. In its severest forms, sustained stress can actually cause brain damage. But you might be surprised to know that researchers initially had a hard time defining the bad agent. There was a thicket of confounders that had to be macheted first.

One prominent confounder was the fact that not all stress is bad for the brain. Mild stress can even boost performance under certain conditions, something behaviorists call *eustress*. We also know that the experience of stress is remarkably subjective. Some people love bungee jumping, consider it exciting and overstuffed with eustress. For others, bungee jumping is their worst nightmare. They get tense just thinking about it.

Even our biology jumps on the ambiguity bandwagon. Suppose you showed me two physiological workups—one from a person experiencing extreme pleasure, the other experiencing extreme stress—then asked me to identify which was which. You'd probably be frustrated with my answer, because I wouldn't be able to tell the difference. Neither would anybody else. They're actually too similar to discriminate.

Research has revealed that many of us were looking at the wrong thing. It wasn't the presence of an aversive stimulus that caused all the bad stuff. It was the inability to *control* the aversive stimulus that caused all the behavioral disfigurement. If you felt control over something obnoxious, you might not even report it as stressful. The more you felt out of control over an aversive stimulus, however, the more likely you were to experience *harmful* stress.

This lack of control can be measured in two dimensions: the inability to control the frequency of the stressor coming at you, and the inability to control the severity of the stressor once it arrives at your doorstep. It's as simple as being asked to do a job for which you will be evaluated without being given a budget or personnel to successfully execute it.

Now we know that out-of-control stress can hurt cognitive ability in virtually every way you can measure cognitive ability. Working memory (short-term memory) is inhibited, as is mood control. Long-term memory formation is obstructed. Stress impairs fluid intelligence, problem-solving skills, and pattern-matching ability. The severest form of this unbridled tension—where *nothing* you do seems to stop the bad stuff from happening—is called *learned helplessness*. It's got powerful enough arms to push some people into the abyss of clinical depression.

Here's an example of someone who is knocking on the doors of learned helplessness. This person wrote a confessional on a now-defunct website for new parents. Things were not going well:

> *I miss having friends. My hubby works nights and I miss*
> *having someone to talk to. I hate working a 9–5 job but*
> *I miss the friendships there, too. I hate that sometimes I*
> *don't play with my child because I'm just trying to have*
> *some me time and I let TV entertain her. I need more ME*
> *time, more husband time and I DO need to spend more*
> *play time with my daughter. I have severe mommy guilt*
> *and I hate it.*

Business and stress

American workers were already feeling the cognitive damage from losing control prior to the pandemic. A survey taken in February of 2020 revealed just how much stress was eating into their work lives: most workers declared they were uncomfortably stressed by their jobs. Those are charitable words, actually. A whopping 61% of Americans felt burned-out, the same toxic feeling that comes from having way too much to do and no time to do it. After the virus hit, that number jumped to 73%.

The reasons for these obviously bad feelings included job insecurity and unmanageable workloads. But the biggest stress source was that American workers felt that the boundaries separating work life and home life were vanishing.

What's relevant here are the on-the-job practical effects of boundaries being blurred. Loss of working memory nearly guarantees you're going to make more mistakes on the job. Loss of mood control means you're going to have more conflicts with your coworkers, your kids—just about everybody. Stretched over time, your mental health becomes at risk, particularly for psychopathologies called *affective disorders*. The two most famous of these disorders are depression and anxiety.

Stress also paints a big target on your physical health, the effects on cardiovascular health probably being the most famous. What's less well known is that stress increases your susceptibility to infectious disease—viral, fungal, and bacterial. We even know why: Elevated stress hormones (cortisol, for example) target specific cells in the human immune system, including a population called T helper cells. As those cells die off, you lose the ability to ward off the microbial bad guys, who, under less stressful situations, you'd defeat just fine. To consider what life is like without T cells, look no further than HIV, the virus that causes AIDS. That virus specifically targets those same T cells, rendering the rest of the immune response impaired. Before successful treatments became available, HIV infections were a guaranteed death sentence.

Quite literally, experiencing work-life imbalances can cripple your immune system's ability to fight even the common cold.

The stars come out

There are certainly nuances to these data. Different occupations felt different levels of burnout, for example. The biggest stress sufferers were people who worked under the Big Tech tent—companies like Cisco—and those working in the "gig" economy—companies like Lyft. But you see it everywhere. Burnout, for most, anesthetizes joy and forces it onto life support.

In varying stages of their lives, people experience the pressures of work-life imbalance differently. Employees in their childbearing years have different worries than people getting ready to retire—and both have different worries from people just starting their careers.

Regardless of job or stage in life, it was—and remains—all about control. This idea has been tested directly, investigating the so-called work-to-family interference path mentioned earlier. Research shows that when employees have more control over their work lifes, it directly affects their families' health. This change in control has measurable short-term benefits.

Here's how we know this: A group of researchers decided to find out what happens to employees if they are granted control over one aspect of their job—their schedule. Test subjects were primarily heterosexual individuals involved in long-term, committed relationships.

The experiment tested a program called STAR (Support, Transform, Achieve, Results), a behavioral intervention protocol designed by a consortium of researchers at the Work, Family, and Health Network. STAR gave employees much more control over how they used their time to balance work with their personal lives. The program additionally instructed their bosses to support—rather than resist—the employee-designed changes. The experiment lasted twelve months.

The end of the reporting period showed that the intervention worked like a son of a gun. Employees who received the STAR intervention became less stressed, experienced decreased burnout, got along better with their bosses, reported increased job satisfaction, and—here's the icing on the cake—had reduced family-to-work conflict. It's the kind of data that inspires a press release as much as a scientific paper.

The program was particularly effective for the women employees in the test group (which is sad but makes a certain amount of sense, as we'll discuss shortly). And there was an unexpected bleed-through effect: giving employees control over their own schedules also worked particularly well in families with teenagers. The adolescents slept better, reported an increased sense of well-being, and had a more positive general affect. I can't think of a greater indicator of a change in a family's emotional dynamics than this extraordinary finding.

Home partnerships

These data remain tasty, but they mostly explore a one-way street, the work-to-family interference route. What about the other direction, the so-called family-to-work interference avenue? Does what happens at home affect work too?

The answer is yes, and it can be intuitively illustrated with a simple question: "Did you sleep well last night?" Your answer will tell me—and any other brain scientist in the room—how productive you're likely to be at work today. If you slept poorly at home, then went to work, you're dragging your sleep schedule behind you like an uncooperative dog. This is just one example of the *giant* relationship between what happens to your brain at home and what shape it's in when work commences. Researchers have looked at couples with and without children to detail the linkage, mostly studying, once again, heterosexual relationships in long-term, committed relationships (primarily marriage).

Let's start by examining domestic spousal relationships, specifically the quality of the marriage. There's both good news and bad news to share. The good news is that having a spouse with a positive disposition—providing insight and stability—increases a worker's productivity.

One paper, examining the personality traits of over 5,000 people, was delightfully entitled, "The Long Reach of One's Spouse: Spouses' Personality Influences Occupational Success." These researchers found that the most conscientious spouses had partners who were the most successful at work, male or female. These lucky partners liked their jobs better, were more likely to be promoted, and not surprisingly, made more money. The data echo a quote from Facebook's COO Sheryl Sandberg:

> The single most important career decision that a woman
> makes is whether she will have a life partner and who
> that partner is.

Consider this example, once again taken from that now-defunct confessional website for new parents:

> I am 37 weeks 2 days pregnant with identical twin
> girls—I'm being admitted tomorrow to deliver. Yesterday
> my dear husband sent my BFF and me to a spa for the
> works. I was given a new cut and highlights, massage,
> facial, mani, and pedi. When I got home, I found he
> prepared every single food I've craved [during] this
> pregnancy for dinner. Tonight he … sent me to bed
> to rest while he readies the house for the babies.
> God I love that man!

The flip side of this research concerns divorce and workplace productivity. As you might expect, it's not a pretty picture.

For most employees, dealing with a divorce alters behaviors essential to productivity. The hot spot appears to be the six months prior to final separation. During that time, the employee may have trouble focusing. They may have trouble remembering things, from making appointments to making reports. Skipping work to attend an attorney meeting or court appearances is common, resulting in repeated absenteeism. The net result is that workers enduring a divorce are 40% less productive than workers in stable relationships, costing American businesses north of $300 billion a year. The bad news is that work life is tethered to home life like two prisoners in a chain gang.

Family matters: the bad news

So, what happens at home doesn't stay at home.

These effects are compounded when children are folded into the mix. Historically, certain businesses have recoiled at hiring employees with families, especially young hires with young babies. One man confided to me he never hires women of reproductive age. "They always leave when they get pregnant," he groused.

That might seem horrifically misogynistic (it is) and illegal (discriminating against a pregnant person is a violation of a federal statute), but it's also breathtakingly shortsighted. Without families, noisy babies in tow, the economies that businesses depend upon for long-term survival would crash and burn. That fact is actually quantifiable, which we'll discuss shortly. Companies that make decisions based on the assumption that employees with families are short-term liabilities rather than long-term assets are willfully valuing the weather while deliberately ignoring the climate.

It's easy to understand why businesses sometimes treat employees with starter families as liabilities. Work priorities suddenly have active competition with what goes on at home. Pint-sized babies invariably introduce giant-sized stressors into a rookie parent's life, which means what goes on at home can be debilitating. This includes a loss of sleep, an upending of a reliable schedule, larger financial costs, a drastic change in daily expectations, and always more work than anyone expects. The volatile out-of-control feelings fueling the engines of burnout roar to life in the first months of parenthood. And the noise doesn't substantially change for years. Though it doesn't have to, productivity at the office usually suffers.

This stress has been quantified. Consider the aforementioned sleep loss. During the first six months of a baby's life, parents average only about two hours of uninterrupted sleep per night. About 30% of new parents fall asleep at work. (Nearly 21% fall asleep in their cars!) The deprivation is long-lasting. A typical mother doesn't return to her pre-baby sleep routine until her child is six years old.

As you might expect, the impact is expensive. Sleep deprivation of all types costs the American economy $411 billion a year. Raising kids, the world's most difficult amateur sport, is hard work for just about everybody. It's an equal opportunity offender, causing relational fractures both at work and home. Some business professionals, at least in the short-term, would just rather not be bothered.

One of the hidden stressors of new parenting isn't the product of newborn interactions but spousal interactions. Research shows that marital conflict increases by a whopping 40% after couples have their first baby. Two-thirds of married couples report a drop in the quality of their relationships three years into their parenting journey.

The occupational effects of having children are especially hard on women in the workforce. Google found that the number of females who quit their jobs after giving birth was twice the rate of the company's average attrition—and that was a pre-pandemic number.

In the post-pandemic world, industries where the work is primarily done by women (education, food service, retail) took big hits. Occupational surrender—which resulted in women leaving the workforce—became so disproportionately common along gender lines during the viral infection, it was christened a "she-cession." The most common reason was that families couldn't afford child-care: somebody had to come home. That somebody became Mom. In December 2020, the pandemic in full throttle, women lost

156,000 jobs. Men, on the other hand, actually *gained* employment, increasing their employment rolls by 16,000.

This disproportionality—and subsequent stress—could be seen even in two-career households during the pandemic lockdown. The University College of London examined the household chore and childcare assignments in heterosexual marriages during quarantine. Women *still* did twice as much housework and childcare as men, though both were home and the chores could presumably be distributed more equitably.

This is a big, hot, steaming pile of stress. It's easy to see why businesses might not see much short-term upside when it comes to families and work.

Family matters: the good news

I do not mean to be down on families, especially for those of you considering becoming parents. Having reared two now-grown sons, I can tell you in retrospect that parenting has been the most heart-warming, heart-pounding, heart-exhilarating experience of my life—and sometimes the most hysterical one, too. Consider these two final stories from that decommissioned parenting confessional website quoted previously, which illustrates this ambiguity nicely.

This from a young family:

> *Currently LMAO. My dear husband is playing "tea party"*
> *with our fou-year-old daughter. Her rules for this party:*
> *Daddy had to wear the pretty boa, and the clip-*
> *on earrings. She is also giving him a lecture in proper*
> *pinky placement (pinkies out). LOL- I'm so*
> *getting my camera!!*

This from an older mom:

> *My daughter is a psychology asst. professor now!*
> *I went to one of her classes yesterday and sat in the back,*
> *and I have an entirely new respect for her as a woman.*

*She is more than just my daughter now. She is a woman
with a passion, and who is educated, and who I can learn
from. I can't believe that is my daughter!*

The power behind allowing employees to undergo experiences like these cannot be overstated. Businesses choosing to see family-raising employees as assets rather than liabilities reap more solid benefits—they're just harder to see at first. But research shows that by altering things at home, especially when families are involved, long-term benefits accumulate at work. The issue is getting management comfortable signing up for marathons rather than always choosing sprints. The single best way to run this race? Insert parental leave into the benefits package.

The results are striking. Providing maternal leave keeps valuable female executives from leaving, with savings in the low six-figure range. (It can cost up to $213,000 on average to replace a top-tier talent.) Google is emblematic of this finding. As you recall, they discovered the attrition rate of women exiting work after child-birth was twice their normal average. Google then introduced a paid maternity leave program. The two-times difference disappeared into thin air. Providing maternal leave for married couples also reduces their overall divorce rates, a potential $300 billion savings.

Paternity leave has also been studied. Similar positive effects have been observed, ranging from a reduction in those aforementioned divorce rates to improving the health of the postpartum mother (possibly because there's now someone local to share the load). This assistance continues past birth. Fathers who experience paternity leave end up being more involved in raising their children, an uptick still measurable years later.

What's so striking about these data is the net cost of paid leave to business. *It's actually zero.* Providing childcare benefits to every employee might sound expensive. Initially it is. The costs of providing programs over the long term, however, reveal an offset.

Whether measuring performance or profitability, the price of such programs roughly equals the cost of employee turnover due to pregnancy or childcare. Hard to believe? Such analysis has been run in multiple companies, notably in California. Once again, Google provides an example.

Here's a quote from Laszlo Bock, former senior VP of "people operations" (HR), talking about their two-month paid maternity leave program:

> When we eventually did the math, it turned out this program cost nothing. The cost of having a mom out of the office for an extra couple of months was more than offset by the value of retaining her expertise and avoiding the cost of finding and training a new hire.

Baby benefits

These data should be compelling enough for companies to sign up for the marathon experience, adopting a robust parental leave program to benefit their long-term bottom lines. But beyond a single company's self-interest, there's an additional, more powerful reason to advocate for these programs. It has to do with something *every* business needs in order to survive decades into the future. I am talking about culture—specifically, the social environments in which children turn into adults and, eventually, into employees.

There's a whole branch of developmental brain science that studies the influence of social stability on a child's long-term brain health. And here, a singular pattern emerges, which is rare for the field. This is such a big thing to say that we need to spend some time discussing how brains develop in the first years of life.

The first thousand days of a baby's life are so important, they influence that baby's behavioral outcomes years later. From learning to talk with others to learning to bond with them, many of the social skills that babies will use for the rest of their lives begin to form in those years. How well they form can even predict the quality of employee they are likely to become when they enter the workforce.

The buckets of data supporting these arguments come from behavioral, neurobiological, and even economic sources. One delightful example originated from the laboratory of researcher Ed Tronick. Many years ago, he showed the importance of a behavior he calls "interaction synchrony." This is a thoughtful form of parent-child communication where the parent learns to assess whether or not their pride-and-joy wants more or less interaction. By paying close attention to the baby's cues, the parent discovers when he or she is (a) overstimulating the child, in which case the parent withdraws momentarily or (b) understimulating the child, in which case the parent pours it on. Once learned, this lovely ping-pong match can be repeated throughout the day.

Characterizing this synchronous choreography is not particularly groundbreaking; many parents have been doing it for centuries. It's also not terribly time-consuming, though it does require a parent to be with their infant throughout the day. What was new about Tronick's work was his discovery of how important this synchronicity is to the rest of the child's developmental program. Said Tronick:

> *The emotional expressions of the infant and the caretaker function to allow them to mutually regulate their interactions. Indeed, it appears that a major determinant of children's development is related to the operation of this communication system.*

There are neurobiological data that support why Tronick—and so many other scientists—place such emphasis on the early years of life. Existing brain cells (neurons) start forming synaptic connections with other neurons at a dizzying rate in those years. In the first twelve months alone, synaptic growth increases more than tenfold. By age three, a single neuron carries an average of fifteen thousand connections.

It's an uneven growth, however, with regions behind the forehead (prefrontal cortex) doing the lion's share of the initial networking. There's also a period in the early years when there's too much synaptic exuberance, and some of these connections get pruned back. Exactly how this extraordinary burst—and pruning—relates to the behavioral milestones isn't well established, though the fact of its primacy is not in dispute.

All this smiling and cooing and neural remodeling has measurable economic impact, a finding rare in my field. Paying attention to kids at their earliest ages has a surprising financial influence on any country that pays attention to those years. The reason? Such attention is especially potent for developing skills of an old friend of ours, executive function (both the cognitive and emotional regulation wings of the party). Two gigantic longitudinal studies commissioned in the early 1970s, often couched as the ABC/CARE studies, demonstrated empirically the benefits on executive function.

Originally commissioned in North Carolina, the principal investigators asked an intriguing question: If you paid attention to the earliest years of development in children born into high-risk, economically disadvantaged situations, how would the kids turn out thirty years later? In the experimental group, disadvantaged children were given an enriched, early childhood learning program. The intervention started when the kids were eight weeks old and lasted until they were five years old. Teams of researchers—actually *generations* of researchers—continued to study the program's impact during the next three decades.

The outcomes were extraordinary. Those kids whose brains had undergone the intervention were less likely to commit crimes, become pregnant as teens, or suffer from substance abuse. They were more likely to graduate from high school and college and enter adulthood with a marketable skill. As a result, they earned more money, were more likely to own their own home, and were more likely to engage in their community as adults. In short, they became everything you'd want from a well-adjusted citizen. The controls, to put it charitably, were not in the same shape.

The results have been analyzed numerous times, most famously by Nobel Prize–winning economist James Heckman. Heckman discovered that, compared with the costs of the program, the rate of return on the investment worked out to be 10–13% per annum per child. He calculated that an $8,000 investment at birth (in 2010 compounding interest dollars) gave rise to a return 100 times the value of the initial input over the life of the person ($789,395). He then went further, ratifying the developmental science. Said Heckman:

> *The data speaks for itself ... Investing in the continuum*
> *of learning from birth to age five not only impacts each*
> *child, but it also strengthens our country's workforce*
> *today and prepares future generations to be competitive*
> *in the global economy tomorrow.*

The importance of birth rates

There is another large-picture reason why companies should place value on employees starting families. Economists and captains of industry react with alarm when their country's population experiences a decline. That currently describes nearly every developed country in the world, by the way. The United States, for instance, is having 300,000 fewer babies per year than it should, resulting in an annual 8% drop.

Why are birth rates so important for long-term thinking, and why are their declines so concerning? It's a story filled with interlocking pieces, and I should probably make something clear before we navigate its twists and turns: I'm no economist. My expertise is the genetics of psychiatric disorders. Where my experience has touched on economic issues is in understanding the relationship between economic trauma (such as financial depressions) and brain function (such as clinical depressions). Exploring that relationship has allowed me to brush shoulders with economist-types from time to time. Here, in a nutshell, is how they've explained their worries about birthrates to me.

1. *Having fewer babies means having fewer workers to fuel a country's economic engines. That results in labor shortages, which, at a minimum, translates into slower growth. There just aren't enough people to do the all the work.*

2. *Fewer people of working age means fewer people buying things. Slower growth and less buying have many negative economic effects. One of the most important is reducing the amount of tax money governments can collect from its citizens.*

3. *This is particularly concerning because older people are living longer than ever. In 1900, for example, the average American died at age forty-nine. By 2015, modern science had boosted that number to seventy-eight. While I consider that change to be a net positive, it's not a monolithically upbeat gift in light of reason #4.*

4. *These older populations are not actively generating income. Yet they're still actively incurring expenses, much of it underwritten by federal dollars (Social Security, Medicare, and Medicaid come to mind). It's a perfect storm: increasing burdens on federal programs at the very time the ability to fund them is shrinking.*

I can see my economist colleagues' concern. These trends are hard to hear about and, for people like me, even harder to write about. At the time I penned the paragraphs in this chapter, I was already sixty-five years old. Which, my economist colleagues are quick to remind me, is the fastest-growing age group in the country.

Living messages

I still like giving and receiving greeting cards—the honest-to-god dead-tree ones. I especially love the ones congratulating someone on the birth of their first child. Some cards are hysterical: "Just so you know, friend, having children is like living in a frat house. Nobody sleeps, everything's broken, and there's a lot of throwing up." Some are more practical but still funny: "Before I had children, I had no idea I could ruin somebody's life by asking them to put pants on."

My favorite ones are never tinged with humor, though. They're ones that balance the profundity of raising someone with the toil it takes to do so: "A baby will make love stronger, days shorter, nights longer, bankroll smaller, home happier, clothes shabbier, the past forgotten, and the future worth living for."

The one that particularly sticks with me was not one I gave, however, but one I received when our eldest son, Josh, was born. The front was the standard congratulations. On the back, the sender wrote in her own hand a quote from John F. Kennedy: "Children are the living messages we send to a time we will not see."

For all the data we examined in this chapter, I cannot think of a better argument than the aforementioned quote for placing the utmost value on the raising of our children. Creating work conditions where families can thrive may be one of the most important long-term social contributions a business can make. To do so is in the best interest of businesses; it is also in the best interest of just about everybody. Businesses need healthy birth rates. Businesses need a future generation that has been raised by caregivers who have the time and the mental and emotional bandwidth to nurture their children's development. Businesses need to consider families as long-term investments.

Progress

Behind this call to action is a badge of shame. Though much of the brain-and-behavioral data we've examined come from American laboratories, the United States is still the only developed, industrialized country without any federally subsidized (and paid for) parental leave for every worker. Not maternity leave, not paternity leave.

Nothing. The fact that parental leave can be such a polarizing political issue almost guarantees we won't have it any time soon. That's always surprising to me and to many of my colleagues. The importance of the early years is not a political issue. It is simply a biological fact.

Happily, the federal government keeps trying things on this front. In October 2020, the government passed the FEPLA (the Federal Employee Paid Leave Act). This law provides twelve weeks of paid parental leave for certain categories of federal civilian workers under certain conditions. It's a nice try, but it doesn't apply to everybody in the country. It doesn't even apply to everybody in the federal government.

Employers in the private sector have made similar attempts, actions accelerated by COVID-19, and the efforts at first blush seemed to show real progress. Even prior to the pandemic, an estimated 70% of firms in the "knowledge business" had already introduced something akin to Google's program (Microsoft, IBM, Reddit, and Amazon, for example). But this promising news is tempered by a sobering reality: they're actually the exceptions. One firm estimated the total number of US companies offering family leave benefits at 6%. A more recent survey disputed this, however, placing the number at 16%. (And still *another* survey claimed the number was closer to 55%.) This statistical turbulence illustrates the current state of flux on this issue, made worse when considering pandemic-related recovery matters.

Regardless of the actual figure, a broad swath of the business community still ignores the issue altogether. Even for those companies who have embraced family leave, their benefits vary wildly in terms of amount of allowable time off and financial compensation during the absence.

I'm no federal policy expert, but I am a brain scientist. If the United States wants to remain competitive in a world where brain capital is the *actual* reserve currency, it has to take care of those brains everywhere it sees them. That begins inside the skulls of our youngest citizens, under the loving care of people who, probably just like you and me, want the chance to raise them to their full potential—and not have to fight constantly with their spouses and bosses while trying to do so.

What to do next Monday

I'd suggest you begin by rewatching the movie I mentioned at the start of this chapter, 9 *to* 5. Parton and her fictive colleagues actually added some of the revolutionary workplace features mentioned here, such as control over working hours and childcare options. Their managers discovered that those changes were responsible for a large productivity boost, something the nonfictive research literature would demonstrate years later. (The movie even predicted that pleas for equal pay would go unheeded, a prophecy that sadly has also proven true.)

As a second piece of advice, I suggest you carefully study the STAR program (references provided at brainrules.net/references), then consider applying its tenets to your own work. Granted flexible work schedules are not revolutionary ideas—it was 1980 when 9 *to* 5 mentioned the concept, after all!—and STAR has the luxury of being underwritten by some pretty serious behavioral science. Feeling like you have some control is the key to managing stress from any source, including the strain generated by your schedule. STAR has the additional benefit of having been analyzed to see if it works, which, as we have seen, it does quite well.

Finally, I'd suggest you further study the Heckman report, the analysis of that thirty-year longitudinal research showing the importance of the early years of life. Follow that with a chaser of developmental neuroscience, using the references cited in this chapter. Then, if you're (a) looking for a job and (b) thinking of starting a family, then (c) consider seriously whether that company has a paid parental leave policy. If they don't have one, don't take the job.

If you're in a company that doesn't have such a policy, I suggest handing the powers-that-be this chapter, or at least a few of the references that support this chapter.

If you're feeling especially ambitious, write or call your elected lawmakers and urge them to support laws that expand the scope of parental leave. Explain that this isn't a political issue, but a practical issue, using evidence from this chapter to support your claim.

Dolly Parton's character offered a toast to her colleagues in the movie's final scene, thinking their innovative workplace changes were the beginning of something new. It would be sad, at the end of the day, to tell her that they still are.

WORK-LIFE BALANCE

Brain Rule: *You don't have a "work brain" and a "home brain." You have a single brain functioning in two places.*

- Stress isn't the experience of averse stimuli; it's the experience of not being able to *control* averse stimuli.
- Feeling control over your work life (for example, your schedule) increases your chances of having a healthier family life. Similarly, having a healthy home life (for example, a supportive spouse) can improve your work productivity.
- While experiencing divorce, workers are 40% less productive than workers in stable relationships.
- When families decide to have children, women are disproportionally affected in the workplace. They quit their jobs more often and, during the pandemic, did twice as much housework and childcare as men, even if both were home.
- To reduce divorce rates of all employees and turnover of female employees, companies would do well to have a robust parental leave program. The long-term net cost to provide parental leave programs is zero.
- To strengthen the future workforce and its accompanying economy, companies would do well to invest time and resources to allow for child development, especially for children from birth to age five.

change

Brain Rule:
*Change won't happen out of determination
and patience alone.*

THEY DON'T EVEN CALL THEM "steam shovels" anymore. It wasn't a great name anyway. They were never really shovels, and they're no longer powered by steam. Now you can call them antique. Steam shovels became obsolete in the years before World War II. This obsolescence so moved author and artist Virginia Lee Burton that she wrote a short children's book about it. She entitled her fictional work *Mike Mulligan and His Steam Shovel*. The book is really about change, which is also the subject of this chapter.

In the book, a character named Mike lovingly christens his steam shovel "Mary Anne". For years they enjoy a happy, lucrative partnership. In their prime, Mike and Mary Anne dig canals and highways and cellars for large office buildings, but this doesn't last forever. Mary Anne eventually grows out-of-date, replaced by more modern models, such as gas and electric shovels. Mike and his steam-powered work wife soon run out of paying jobs.

Mike and Mary Anne are very sad until, one day, they discover that the nearby town of Popperville is building a new city hall. Excited, Mike tells a town official that he and his talented steam machine could dig the cellar for the city hall in one day. With the townspeople cheering him on, that's exactly what Mike does.

Now, you'd think that would be the happy ending of this tale. There is a cheery finish, certainly, but not before Mike and his steam shovel are confronted with one last problem. In his haste, Mike neglects to build a ramp that would allow the steam shovel to leave the pit after they finish digging: Mary Anne is stuck. Burton drew a picture of a forlorn old technology, sitting in the middle of the hole they'd dug, with no way out.

It's a perfect picture for the consequences of failing to roll with the times.

The book was published at the end of the Great Depression, a time when economic turbulence raked across the American workforce like a hurricane. Mike's forced adaptation wasn't unusual in those days. It's hardly unusual today, especially as we contend with the whirlwind of another economic disruption, this one of viral origin. In this, our final chapter, we're going to discuss the central lesson behind *Mike Mulligan and His Steam Shovel*, one that's been perhaps the most reliable constant during these last eighty years: change is hard, necessary, inevitable, and, for those not willing to adapt, trapping you in a hole of your own making.

In this chapter, we'll discover that (a) the brain really does hate to change its mind and (b) there are ways to make change easier on our poor bag of nerves. We'll cover the research domains concerning change adaptation, which curiously comes mostly from investigations into habit formation. We'll describe how (and why) we create habits, both bad and good, and what it takes to turn the former into the latter.

I'm happy to report that the research on change includes a large dollop of optimism. It also includes a gentle warning. As Helen Keller, a contemporary of Burton, once noted:

> *A bend in the road is not the end of the road ...*
> *unless you fail to make the turn.*

Change is hard

Change is ridiculously hard for people, even when the change results in positive outcomes. Consider what happened to commuters when London Underground workers went on strike in 2014, closing several "tube stops."

The strike severely disrupted travel, forcing customers to find alternate routes. Surprisingly, the new routes often saved the commuters time. (A detoured worker saved an average of seven minutes on their normal thirty-minute ride.) Despite these extraordinary numbers, only 5% of the subway riders permanently changed their commuting habits once the strike ended. A whopping 95% reverted back to their older, time-consuming habits.

Change resistance can affect medical outcomes too. An astonishing 91% of people who undergo emergency bypass surgery go back to their unhealthy pre-surgery lifestyles, despite admonitions from their doctors that they will die if they don't change. Worldwide, more than half the people in the industrialized world diagnosed with severe medical conditions *of any type* do not take the prescribed medications that could literally save their lives.

Change resistance, it seems, is everywhere.

Change is equally hard in business settings, though quantifying exactly how hard hasn't been easy. Some studies claim that 70% of all business change initiatives fall flat on their enthusiastic faces, and that this depressing figure has been stable for decades. Other studies claim this is statistical bird poop, that the real figure is about 10%, with mixed success hovering around 60%.

Why the disparity? Part of the problem lies in defining exactly what you mean by change and what you mean by resistance, then getting everybody to agree that your definitions are one-size-fits-all.

Good luck with that.

The best definitions of change probably reflect lived experience, which almost always involves conceptualizing a continuum. At a minimum, change can be defined as a disruption. On one side of this definition are small, incremental evolutionary adaptations leaving the status quo's framework intact. They can be annoying as pinpricks, but not particularly life-altering. On the other side lie the truly revolutionary alterations, giant transformations that shatter the status quo's framework. They're as concerning as heart attacks, and absolutely as life-altering.

Resistance is an equally important factor behind the statistical turbulence. One can generally define "resistance to change" as anything striving to protect the structure of the status quo. There are many types of resistance too. Some resistances are large and openly rebellious, sometimes resulting in lawsuits, relational fractures, or in the case of geopolitics, armed conflicts. Some resistances are active, but small and incremental, creating barely noticeable obstructions. Other resistances are passive, functioning essentially as weaponized inertia.

What they all have in common is that change is hard, regardless of how you define *change*—and regardless of how you define *hard*.

Why change is hard

However you characterize resistance to change, the disruptions are almost universally hated by anyone over the age of, say, three days old. And we think we know why.

Human beings, it turns out, are control freaks—a phenomenon so important, it burrows into our definition of *stress*. You might recall from a couple chapters back that stress doesn't bother us as much as our inability to feel in *control* of it does.

People can be particularly anxious when thinking about the future. The anxiety is initiated by a cognitive gadget called *mental time travel* (MTT), a processing feature developed ages ago. Mental time travel is the ability to envision the consequences of future decisions based on present actions. It's a card-carrying member of that famous behavioral suite we've been discussing on nearly every other page, executive function.

How does MTT relate to change resistance? When people are asked to change, they risk ceding control of their future. They may try to imagine life with the change in place, envisioning upsides and downsides, using the gadget to predict and thus control what happens next. The effort causes many people discomfort. Trying new things is risky, after all. A change may cause pain. An adjustment may make things worse. Since "new" might equal pain, "new" might equal "bad." Nobody I know likes painful disruption.

MTT is not the only force deployed when we react to change, however. At the same moment we're thinking of "the new," we're making continuous comparisons with "the now." The present is a land where control isn't much of an issue. It's familiar, obvious, and—compared with the uncertain future—maybe even comforting. Our rationale is that if no change equals "less pain," then it also equals "more good."

The asymmetric valuing of these two perceptions—the intangible future and the predictable present—forms the basis of change resistance in the human brain.

If you do science for a living, you get used to a fair amount of change. My journey into disruption occurred long before I put on a lab coat, however. I'm old enough to have written my high school essays using the same technology familiar to fifteenth-century monks: ink and paper. I remember the day the mighty word processor came to town, banging on the walls of my medieval status quo, announcing it was time to leave the Middle Ages. When I was in college, a very early version of Microsoft Word overwhelmed me with the belligerence of an occupying foreign power.

At first, I resisted the change, and with a mighty fury. I couldn't even *type*, for heaven's sake. Nothing was familiar or comfortable using a word processor. White paper had been maddeningly transformed into a deep-blue screen. The letters on a "page" were no longer made of dark ink, but tiny points of light. Prose was typed in staccato bursts—the sound of which reminded me of machine guns—rather than the fluid, languid rhythm of cursive inscription.

It probably took me half a year to make the switch. And I hated every minute of it. I kept bumping into an efficiency argument. Papers were due. Grants needed writing. Words I could pen with my hand in fractions of seconds now took painful minutes as I tried to type, my fingers hunting and pecking like chickens.

Why did I hate this? This newfangled way of doing things came with upfront costs. Rewards, if there were any, lay far in the future. Even with our fancy MTT ability, we're still not very good at understanding the long-term consequences of short-term behaviors.

X and C systems

Researchers have attempted to understand the brain networks behind these perceptions. Matt Lieberman et al. think they've found two of them. They call the first the "X-system," originating from the word *refleXive*. These groups of neurons are fast, efficient reactors to specific stimuli felt in real time. They have twin preoccupations: (a) processing immediate goals of almost any kind and (b) contrasting how these activities compare with past experiences, especially previously formed beliefs and habits.

The second system is called the "C-system," originating from the word *refleCtive*, and acts like X's wiser, older brother. It is constantly advising, correcting, and disputing the conclusions of its neural sibling. The C-system is not automatic. It reacts to stimuli more slowly, sucking up tons of energy in the process, to perform its supervisory operations. If you embrace the need for change and persevere—even though the initial costs are high (I'm looking at you, Microsoft Word)—you can probably blame your C-system.

Not everybody agrees with this neural taxonomy, of course. But it has the luxury of presenting testable ideas. Researchers have made real progress mapping these behaviors to specific brain regions, for example. We now know more than ever before about how the brain reacts to—and sometimes embraces—change, even if it's the last thing it wants to do.

Neuroanatomy

One of the most important structures involved in the X-system is something called the *basal ganglia*. It's a fairly big region, with lots of nonmoving parts. The basal ganglia looks something like a large-headed comma curled up in the middle of your brain.

We used to think this neurological punctuation mark was involved primarily in motor functions. Now we know the basal ganglia has a lot of side hustles. One is involved in the creation of habits. Another is generating reflexes. Reflex generation and habit creation often involve motor skills, but the various regions of the basal ganglia become activated for any familiar, repeated, and ultimately automatic action. Ever gotten home after driving from work, having completely forgotten how you got there? Blame it on your basal ganglia.

The C-system also engages a wide variety of neural substrates. The biggest are the energy-guzzling regions that mediate executive function, the regions just behind your forehead. This makes sense. Recall that executive function is involved in impulse control, something that gets challenged when confronting change. Your natural impulse may be to run away from the new thing you're supposed to do (like my word processor nemesis). Executive function reminds you to stay the course, regardless of how you feel about it. To say that such behavior takes an enormous amount of fuel is an understatement; it's something to which the brain is nearly allergic. No wonder we resist change: to the brain, it's a darned waste of energy.

The X- and C-systems aren't the only neural networks the brain uses when reacting to change. One of the most interesting involves the so-called *error detection system*. This system is deeply concerned with expectation management.

What do I mean by expectation management? Consider, for example, what would happen if I gave you a bottle of Chanel No. 5 perfume and asked you to smell it. Unbeknownst to you, the bottle had been previously spiked with butyric acid, a chemical that smells like vomit. As you take a whiff, you are suddenly horrified by this *eau de puke*. The reason you react? Your error detection system immediately sent out an all-points bulletin to the rest of the brain that expectation and actuality weren't in sync. When it detected butyric acid, your entire pattern-matching system climbed to DEFCON 1. The red alert this activation causes is uncomfortable for us, and may be one of the reasons why change is so hard. As a survival mechanism in the wild, the alert is indispensable. As a reaction to a rapidly changing corporate environment, maybe not so much.

How long does it take?

Once again, we observe that energy conservation is central to why we do the things we do. And we like to conserve a lot. It's estimated that we relegate 43% of our daily activities to autopilot. The natural question to ask when we want to make a change is: How long does it take for a new habit to form once our brains have decided it's in our best interest?

Unfortunately, we just don't know.

For many years, the magic number was twenty-one days. The figure originally came from 1950s plastic surgeon Max Maltz. He wondered how long it would take for his patients to adjust to their new, surgically augmented bodies post-op. Maltz observed it took twenty-one days. He wrote a book about this, inexplicably entitled *Psycho-Cybernetics*, which eventually sold 30 million copies. Soon, twenty-one days became the universal benchmark for how long it takes to change nonhabitual to habitual.

Not everybody was convinced people could change entrenched habits in three short weeks, however. Researchers from Europe added some rigor to the question years later, introducing novel personal daily routines not involving face-lifts. Randomly assigned research subjects created a new habit, then documented how long it took for it to become reflexive. The numbers fluctuated as wildly as the stock market in a recession. Some people took eighteen days to ink the new deal. Others took 254.

The answer to how long it takes to form a new habit is simple, if not frustrating: though it appears to be variable, two to three months is now considered the benchmark. Even that number is hardly one-size-fits-all.

Such unevenness is observed when examining efforts to change the behaviors of groups too. The effort to stop smoking has been touted as a social success, at least in the United States. The CDC found that one federally funded program inspired more than 2 million people to quit.

But group campaigns don't always work. Researcher Wendy Wood cites a specific failure of California's famous "5 A Day for Better Health" program, designed to get people living in the Golden State to eat more fruits and veggies. At the beginning of this let's-get-smart-about-our-diet campaign, only 11% were practicing this healthy habit. Five years and many millions of dollars later, the number adopting this healthy habit was—cue the frustration—11%.

Awareness wasn't the problem, curiously. At the campaign's beginning, a paltry 8% of the populace knew 5 A Day was a good idea. By five years later, that figure increased to 30%. The awareness didn't change anybody's behavior, though. It seemed to change only people's guilt about what they didn't eat.

Two misconceptions about the failure to change

Why do some programs fail to change while others succeed? Why do some *people* fail to change while others succeed? Many reasons exist for the turbulence, with explanations as varied as the programs—and the people—themselves. Two misconceptions almost always make guest appearances when researchers try to explain the failures. Both concern the expectations participants bring to the effort.

1. *Change happens if I have the determination to make it happen.*

The first expectation involves one of the prickliest compound words in the English language: willpower. It's been the subject of intense study, one of the most famous of which involves videos of humans trying to resist temptation. The videos are both excruciating to watch and beyond funny. They almost always involve preschoolers.

In a typical experiment, a solitary child is seated at a table upon which sits a marshmallow. An adult (who has to leave for a few minutes) offers the child a deal: eat the marshmallow right away or wait until the adult returns—with the mutual understanding that if the marshmallow hasn't been eaten by the time the adult comes back, the child will receive a second confection. The adult then leaves, the camera still running.

Cue the excruciation.

Some kids simply stare at this little white puff of goodness, looking longingly at this now-unsupervised temptation. Some kids look away. One child turned her back to the marshmallow and focused on the blank wall behind her. Another sat on his hands. Another tried closing his eyes and reciting some kind of math lesson.

Most of these strategies were useless, sadly. The majority of the kids eventually picked up the confection, touched it, tasted it, put it down, then popped it in their mouths. They didn't have the ability to resist.

This is an example of the now-controversial marshmallow experiment originally designed by psychologist Walter Mischel. Though other researchers have had difficulty replicating his initial results, the videos themselves don't need much explanation. These kids were wrestling with some aspect of impulse control. Some aspect of *willpower*. Some people believe that programs like the 5 A Day project fail because the participants act like these kids. They simply don't have the discipline to put down the Doritos and pick up the carrots.

Later research showed this couldn't be further from the truth. I will give a more full-throated treatment to that statement after we have examined the second misconception people believe about change.

2. *Change will happen if I am patient enough.*

This one is familiar to anyone who fights the temptation to throw their computer across the room when a website doesn't load fast enough, which appears to be most of us (more than half of all users will actually abandon a site if the delay is longer than three seconds). Patience bridges the anxious distance between the desire for instant gratification and the realization that results may take longer than a few seconds. It's important, because the accumulation of small events extended over a long time pushes us toward many critical life changes, big and small.

I experienced the positive version of this lesson the more hours I accumulated on the word processor. As I got used to cutting and pasting whole blocks of text, I realized I'd no longer need erasers. *For the rest of my life.* That made me really happy. I eventually discovered the joys of saving multiple copies of written text, then comparing the two in real time to see which was better. That made me even happier. I began to embrace the change from ink and pad to keyboard and pixel. It just took six months of small eurekas to see it. If I felt I needed to be instantaneously comfortable with a word processor before using it, I never would have switched.

You'd think, given the power of long-term accumulation, that the advice would be easy to comprehend: Everybody just needs time to allow change to work its magic. Everyone needs to slow down and have patience with the process.

And that's good advice, but it doesn't go far enough. The frustrating variability pops up again: the 5 A Day program went on for *years* but didn't move the behavioral needle at all. Moreover, not all accumulations result in positive outcomes; time isn't always a friend. Divorce is one painful example. Marriages seldom die from sudden relational heart attacks. They almost always hemorrhage to death, usually from small pinpricks accumulated over years of emotional wounding.

Loss of a job due to burnout works this way too. When people "suddenly" quit, it's often from the accretion of what Robert

Sapolsky calls "micro-stressors." Individually, the micro-stressors may not seem like much—they're called *micro* after all—but collectively unfolded over time, they produce a strain that makes people "suddenly" resign. They are small, aversive events stretched over time, like a body on a rack.

Clearly, we need more than just *time* to effect positive change. What is the missing ingredient? If it's not just patience, if it's not just willpower (I will tell you why it isn't later in the chapter, I promise), then what allows for positive long-term conversion to take root?

Believe it or not, we think we know the answer. It's kind of embarrassing: what's missing is convenience.

Friction

For many years I've enjoyed a program—and now podcast—called *Hidden Brain*, broadcast by award-winning journalist Shankar Vedantam. One episode featured one of the subjects of this chapter: habit formation.

Vedantam began his podcast in a Seattle building, one with which I'm quite familiar: the Bullitt Center. It's a six-story office building perched on a hill, with upper stories allowing magnificent views of both downtown Seattle and the distant Puget Sound. The Bullitt Center is a virtual buffet of good ideas. It has been described as the greenest commercial building in the world.

The reason Vedantam began with the Bullitt Center concerned the gobsmacking thing that happens to your mind the moment you reach the reception area. You're immediately confronted with a towering six-story staircase, built from warm Douglas fir, with broad landings, soaring to the building's uppermost floors. As you ascend this staircase, the views of the city and water unfold before you. Your desire isn't to take an elevator to your destination when you come in, but to walk up those stairs.

And that's what people do. Fully two-thirds of people who have a meeting on the top floor arrive at their destination by taking those stairs. It's been appropriately named the "irresistible stairway."

This design was no accident. The architects who created the building made the interior excruciatingly convenient for getting exercise. Perhaps they had read how sedentary office behavior is a health risk for employees. Perhaps, like so many in Seattle, they just liked hiking in the great outdoors and wanted to bring some of the experience inside a building. There certainly are elevators, clearly marked for people who won't or can't take the steps. However, elevators aren't the first thing you observe when you walk in, even if it's the first thing you're used to looking for when you enter multifloored buildings. What you see is everybody doing aerobics.

The reason Vedantam began his podcast with the Bullitt Center was to illustrate a concept called *friction*. It's been championed by researchers like Wood (who was a guest on the show). Friction is a description of environmental forces capable of shaping people's habits. Environments that thwart new habit formation are christened "high friction" spaces: *I won't take the elevators, because it takes too much energy to find them.* Environments that allow, maybe even promote, new habit formation are christened "low friction" spaces: *I will take the stairs, because they're right there and, man, do they look compelling, and besides, all my friends are there.* For most experiences, you can swap out the word *friction* for the word *convenience* and, in the best-case scenario, the words *convenience and delight*.

We'll have more to say about the role delight plays in a few pages. They're the happiest pages in this entire book.

Frictionless retail

As a way to study the impact of convenience on behavioral change, friction research has many admirers—and a wide variety of empirical support. If you live 5 miles away from a gym where you hold a membership, for example, your visits will average once a month. But if you live 3.7 miles away, your frequency jumps to more than five times per month. The closer you to live to the facility, the more likely you are to use it. Less friction equals more compliance.

Food studies have shown the same thing, at a depressingly ridiculous level. If you've a choice between healthy food (a bowl of apples) and unhealthy food (a bowl of buttery popcorn), which one you'll consume is actually a function of which bowl is closest to you. If fattening popcorn is easier to reach than healthy apples, you'll go for the calories (consuming about 150 in one experiment). But if the apples are easier to reach than the popcorn, you'll eat the apples (consuming about 50 calories in the same experiment). In these cases, the friction was related to ease of access, just like the stairs in the Bullitt Center.

Suddenly we have a possible explanation—and solution—for the 5 A Day program. The program designers might have enjoyed more success had they placed produce stands within easy reach of the consumer: fruit carts on every street corner on weekdays, near the entrances and exits of bars on Saturday nights, in front of churches on Sunday mornings.

You can see evidence that retailers are paying attention to friction almost everywhere you go once you know where to look. Grocery stores do their best to keep items they really want you to buy at eye level, for example. You're more likely to purchase them for the simple reason that you don't have to reach too far up—or stoop too far down—to put them in your grocery cart.

The online environment may give us the purest examples. Uber, Airbnb, and Rocket Mortgage (slogan: "Push Button, Get Mortgage.") all have tried to make purchasing their products and services as frictionless as possible. My favorite illustration may be Amazon, a company that legendarily elevated friction-free retail to an art form. My favorite ease-of-use example? The "Buy Now" or "Buy with One Click" button.

And Amazon, as you may have heard, is taking over the world.

Examples and types of friction

Happily, there are many strategies to exploit low friction that don't involve world domination but are still fully capable of leading to sustained behavioral changes.

Building new behaviors on the back of preexisting habits is one such strategy. Many people exploit their bedtime routines for just this purpose. I used to sometimes forget to set the house alarm before going to bed. I got in the habit of taking the alarm remote with me when I went to brush my teeth. As I reached for the toothpaste, I'd see the remote and immediately set the alarm. With time, the new behavior was triggered by a preexisting dental habit. Researchers formally call this behavioral piggybacking *stacking*.

Another is termed *swapping*. This strategy involves taking advantage of some preexisting habit—not to add something, but rather replace something. One person I know used swapping to embrace a healthier lifestyle. She wanted to lose weight and cut down on her caffeine consumption. Since she went to Starbucks every day, she decided to change her order, going from a full-fat triple latte to a half-decaf Americano with no cream. This covered both bases simultaneously, executed in context of a preexisting routine.

If you're smart, you can even do mixtures of the two. Woods, the scientist interviewed on the *Hidden Brain* podcast, gives a terrific hybrid example: She wanted to increase the probability of getting regular exercise. She decided to go for a run first thing when she got up, making exercise a part of her normal morning routine. Classic stacking behavior. But she also employed a swapping strategy, in this case altering the concept of *pajamas*. She decided to go to bed in her gym clothes instead of her normal nightwear. When she got up, it was nearly frictionless for her to slip on her sneakers and go for a run.

These strategies are quite effective at establishing new behaviors, especially ones that need a routine to sustain them. But the news isn't all good. These strategies don't always work for everyone, however frictionless they may make things initially. Some may tire of the new routine. Some may stack too much on preexisting habits. Long-term change is once again a problem.

Fortunately, researchers have a ready rejoinder. You can increase the probability of friction-reducing success, creating lasting behavioral change, if you know a little brain science in advance. It is to that brain science, some of which will sound familiar, that we turn next.

The delight of dopamine

I wrote a few paragraphs ago that I would cover in greater detail the role *delight* plays in effecting behavior change. I'm thrilled to revisit our discussion of dopamine, the happiest neurotransmitter on earth.

Dopamine is manufactured by specific neural networks deep inside the brain. The networks are collectively called the *dopaminergic system*. Plural *systems* would probably be a better word, as there exist at least four dopaminergic subnetworks. They're a talented group, mediating everything from motor functions to rewards and delights. They also play a critical role in the initial moments of new habit formation, especially those networks inhabiting the so-called *mesolimbic system*. New behaviors don't stand a chance of becoming a permanent part of your life if these reward systems are not engaged.

Not just any reward will do. Dopaminergic lollipops need to have three characteristics to function effectively in behavioral modification. One of the reasons why habits fail to form is that we ignore them. Think of these happy systems as finely tuned clocks, needing some very specific requirements to work properly.

Characteristic #1: *The reward needs to be immediate.*

You need to reward yourself immediately when you engage in a behavior you're not used to performing. And I do mean *immediately*.

Delay the reward by more than a minute, and you risk losing dopamine's critical assist.

Why so fast? It comes from the well-established fact that learning always requires the creation of new neural connections. These associations are surprisingly fragile in their first moments of life. We've discovered that dopamine helps supply the superglue that cements these associations into place. But you have to apply the glue quickly before the associations fall apart—they're that delicate. You can't say, "I will do this now and reward myself when I get home tonight," and expect the new behavior to stick. In this need for immediacy, your brain acts something like a two-year-old.

The next characteristic is probably the cruelest of the three, and easily the most thoroughly researched, mostly because it's made a few people quite a pile of money.

Characteristic #2: *The reward needs to be uncertain.*

Reward predictability (or rather, reward *un*predictability) determines whether new behaviors will get tenure in your brain. *Unpredictability* means two things here: uncertainty of frequency and uncertainty of quality.

Frequency concerns the schedule by which the rewards are handed out. Though the one-minute rule is pretty bulletproof, research shows it's better to experience those one-minute rewards at indeterminate, random intervals, rather than on reliable, predictable schedules. You shouldn't always get a reward just because you did something right (though you could make the case that reward frequency should be accelerated near the beginning of your behavioral change, akin to overwatering a plant that has been repotted).

Uncertainty also applies to quality. Rewards are most potent at creating durable change when you're not always sure what value of goody you're going to get. Rewards that are bigger than expected, or different from what's expected, always work better than those that routinely possess the same quality. Turns out your brain likes pleasant surprises as much as you do. Varying the quality strengthens the "stickiness" of the habit you wish to establish.

You don't have to look any further than your neighborhood casino to see these dopamine principles at work. People who design slot machines have known about this uncertainty for years. They know you will form a much more durable coin-feeding habit if you are randomly rewarded, varying the amount and frequency of the goodies each time. The optimal frequency? Keep the probability of winning around 50%.

The third and final characteristic is understood as a contrast between extrinsic and intrinsic rewards. It definitely requires an explanation.

Characteristic #3: *Extrinsic rewards do not work as well as intrinsic rewards in shaping behavior.*

Extrinsic rewards are defined as any reward lying outside the trigger experience, even if stimulated by that experience. I used to reward myself with an ice-cold beer after I'd cut the grass on a hot day, for example. There's no intrinsic relationship between lawn mowers and a cool, crisp lager. It was my reward simply because I like an ice-cold beer on a hot day.

Intrinsic rewards are the opposite and, interestingly, are far more potent at driving change. In my book *Brain Rules for Baby*, I described an intrinsic reward experience that occurred the first time I ever played the video game *Myst*. I fell hard for this old graphic problem-solving adventure antique, filled with a compelling narrative and some of the loveliest digital art I'd ever seen in my life. I was the perfect customer for *Myst*, because the more time I spent with this achingly immersive environment, the more I was rewarded. But not with money or fame or anything remotely public. No, I was rewarded with a new, previously hidden part of the game, which revealed even more beautiful digital art. My interest in the game grew deeper and deeper, and the only thing that drove me was my interest in experiencing its new visual goodies.

This is a perfect example of an intrinsic reward. This type is directly "in the flow" of the event, internal to the effort being supplied, dependent entirely on my own input in the context of experience.

Though most people usually end up rewarding themselves with an admixture of the two, extrinsic and intrinsic rewards are not equal in their power to create new behaviors. As mentioned, rewards that are directly tied to the consequences of an action (think the sense of contentment you feel after doing someone a favor) shape behavior more powerfully than external ones. Turns out *Myst* always trumps cold beer.

Not about willpower

It's time to make good on a promise I made. A while back, I mentioned there was a misconception about the role willpower plays in habit formation, and I pledged to explain this in more detail later. That later is now.

For years, we thought being able to replace one habit with another was simply a matter of spine. We believed, for example, that the reason 85% of people who lost weight gained it all back within five years was that they just were too weak to say "no."

Further research has suggested this explanation is a misconception, or at least not the whole story. Going back to the drawing board, some researchers concluded that chronic recidivism wasn't a matter of weakness, but one of exhaustion. Called *ego depletion*, the idea is that you wake up in the morning carrying only so much willpower, like gas in a tank. When your impulse-resisting fuel gauge goes empty—perhaps you had to fight temptation too much that day—your willpower sputters and dies.

Nice idea, and there's certainly strong evidence for it, but it doesn't explain everything we know about the issue.

One massive study from Germany measured people's self-control using the gold standard for measuring executive function. When the study was conducted, researchers deployed what was, at the time, a sophisticated recording system (using what amounted to beeps and pagers). The participants recorded how many times throughout the day they were tempted to yield to a bad habit, and how often they tried active resistance.

The hypothesis was that people who scored high on this test would not yield very often to whatever daily temptation they faced. That's not what the researchers found. They discovered that people with high scores didn't resist temptation any better than low-scorers. People who scored high simply didn't encounter as many temptations throughout their day. They had constructed their lives in such a fashion that temptation wasn't something they regularly experienced.

The conclusion can leave skid marks in your brain. The data suggested that their lived-in environment was an active participant in temptation resistance. High-resisters didn't regularly need to use their store of willpower petrol. People who kept off the pounds long-term did so because they learned to banish unhealthy foods from their houses altogether.

Sound like a friction issue to you? Perhaps mixed with some ego depletion, contextualized with an environmental twist? You can use friction-full environments to prevent a bad habit just as easily as you can use friction-less environments to create one. It appears that keeping temptation scarce is just as important as resisting temptation once it's encountered.

But does willpower not play any role then? After all, people who keep the weight off are only an internet click away from ordering a pizza, just like the rest of us.

Further research shows that willpower does play a role, but only when used correctly. Willpower is great for accomplishing short-term goals—sprints, as it were—but it is a horrible marathon runner. What succeeds in the long run is environmental reengineering.

You have to create a lifestyle where the temptations aren't easily accessible. You will then have enough self-control to resist them when they do come knocking on your door.

So, if you think you can't change solely because you don't have enough willpower, you're believing a myth. It's much more complicated than just saying "no." If the temptation appears easily and often enough, you've lost the battle before you even knew it began.

What to do next Monday

We started this chapter with a synopsis of the book *Mike Mulligan and His Steam Shovel*. You might recall that Mike and his aging Mary Anne forgot to build an exit ramp from the cellar they were constructing for a new town hall. The machine was stuck in a hole from which she couldn't escape.

What I didn't tell you is that the story concludes when one of the onlookers at the construction site, a bright little boy, suddenly gets an idea. If Mike is willing to repurpose the steam shovel into a furnace, right where the digger currently sat, the problem would be solved. They could just build the town hall over the reconfigured machine. Mike could even become the janitor!

This little boy's suggestion is a terrific example of adapting to change in a way that provides benefits. Most everything described in this chapter was aimed at making such adaptations easier to do.

Time now for some practical advice, which involves memorizing three buckets of knowledge.

Bucket #1: *Remember the power of friction.*

Creating change involves the judicious engineering of your lived experience. Increase environmental friction for habits you want to break. Decrease environmental friction for habits you want to cultivate. Remember, you're less likely to eat that popcorn if it's too hard to reach. You're more likely to go on a morning run if you sleep in gym clothes.

Bucket #2: *Remember the power of rewards.*

Adapting to change involves rewarding yourself in small steps. Begin by creating a list of rewards that (a) are pleasant for you and (b) can be experienced quickly. Give yourself a reward immediately after you've started to change your behavior. Then, as time goes by, switch to an increasingly irregular schedule. If possible, make the reward intrinsic to the activity you're trying to cultivate.

Bucket #3: *Understand the limitations of willpower.*

Effortful self-control is useful in the initial stages of change but has limited utility over the long haul. You should identify as myth the idea that "Willpower can conquer anything that bugs you." You should amend that sentence to say, "Willpower can conquer anything that bugs you, *at first.*"

Putting these ideas into practice is the best way to make the changes you need, both large and small. Who knows? You might end up with a fate like Mike Mulligan and his beloved steam shovel. Barton ends her book this way:

> *Now when you go to Popperville, be sure to go down in the cellar of the new town hall. There they'll be, Mike Mulligan and Mary Anne ... Mike in his rocking chair smoking a pipe, and Mary Anne beside him, warming up the meetings in the new town hall.*

CHANGE

Brain Rule: *Change won't happen out of determination and patience alone.*

- People are resistant to change because they risk ceding control. People tend to believe that "the new" could be worse than "the now."
- Considering and ultimately making a change requires your brain to use a considerable amount of energy.
- To be more successful in forming a new habit, give yourself a reward immediately after practicing your new behavior.
- Increase environmental friction (make it inconvenient) around bad habits that you want to break. Decrease environmental friction (make it convenient) around good habits you want to cultivate.
- To change your habits for the long haul, create a system of friction (for old habits you want to break) and rewards (for new habits you want to start). Willpower on its own has limited utility.

concluding thoughts

We began this book with a thought experiment: How might the brain respond to work environments were those environments designed with the organ in mind? With the twentieth-century lessons of Mike Mulligan safely tucked into our twenty-first century future, we've come to the end of that experiment.

Remember when I stated that gloves are designed with five fingers because our hands have five fingers? Businesses would do well to take such ergonomics into account and fashion the work experience around the cognitive shape of the brain, whether addressing power or creativity or stressful experiences or how to keep people from falling asleep during PowerPoint presentations. I hope you can use the suggestions in this book to tailor your work-place to your brain's natural contours.

It's okay if you can't remember all the details in these pages. Most everything can be boiled down to a single idea anyway: nearly every suggestion involves learning how to become less self-centered. Creating effective teams involves *not* dominating a conversation and *not* interrupting someone when you're not the center of attention. Being an effective leader involves marinating your decisions in a bath of empathy—which means consistently thinking of other people's experiences. Conflict management involves you removing yourself from your own disputes so that you effectively become a third person, a videographer, witnessing the conflict rather than participating in it. The research world has a fancy term for all this egalitarianism: *social decentering*. Reimagining the workplace mostly involves remembering one of the most basic rules your parents taught you: think of others more than yourself.

Won't you be my neighbor?

Fred Rogers, you may have surmised, is a hero of mine. I've one last story to share about him as we conclude. In 1997, during the Emmy Awards, Rogers was given a Lifetime Achievement Award. You might forgive him if he chose to be full of himself that one night, but instead his acceptance speech showed just how socially decentered he was. It took less than three minutes, but it was enough to wring tears from all the hundreds of hardened executives, ambitious TV stars, and overworked production crews who heard it.

"Oh, it's a beautiful night in this neighborhood," Rogers began, nodding to actor Tim Robbins, who had introduced him. "So many people have helped me to come to this night. Some of you are here. Some are far away. Some are even in heaven. All of us have special ones who have loved us into being." The rambunctious adoring audience suddenly quieted.

"Would you just take, along with me, ten seconds to think of the people who have helped you become who you are?" he implored. "Those who have cared about you and wanted what was best for you in life. Ten seconds of silence? I'll watch the time." He stretched out his arm to look at his watch and silently counted down ten seconds. This was enough for the TV camera to gently reveal why everyone had quieted: people's eyes were filling with tears. Some were looking down, wistful; some appeared to be remembering something painful. Everybody's brain was making room for somebody other than themself for about ten seconds. As the time expired, Rogers finished this way:

> Whomever you've been thinking about, how pleased they
> must be to know the difference you feel they've made.
> Special thanks to my family, and friends, my coworkers in
> public broadcasting … for encouraging me, allowing me,
> all these years, to be your neighbor. May God be with you.

As the applause and adulation erupted, some people were still taking out their handkerchiefs.

That short moment, those magic ten seconds, is everything this book has been about.

If you have a chance as we part ways, you should take up Rogers's gratitude challenge. Then you should go to the internet and see the speech for yourself. It's my last suggestion about what to do next Monday, and believe me, it's the best thing you'll do all day.

references

Extensive, notated citations
at www.brainrules.net/references

10 brain rules for work

1.

Teams are more productive, but only if you have the right people.

2.

Your workday might look and feel a little different than before.
Plan accordingly.

3.

The brain developed in the great outdoors.
The organ still thinks it lives there.

4.

Failure should be an option—as long as you learn from it.

5.

Leaders need a whole lot of empathy and a little
willingness to be tough.

6.

Power is like fire. It can cook your food or burn your house down.

7.

Capture your audience's emotion, and you will have their attention
(at least for ten minutes).

8.

Conflicts can be resolved by changing your thought life.
It helps to have a pencil.

9.

You don't have a "work brain" and a "home brain."
You have a single brain functioning in two places.

10.

Change won't happen out of determination and patience alone.

acknowledgments

Warm appreciation goes to my editor, Erik Evenson. Thanks for so many insightful ideas, lively discussions, and unflagging optimism. You were so much fun to work with!

Additional thanks goes to Stephen Branstetter, Tim Jenkins, Ryan Mecklenberg, Margot Kahn Case, Katie Prince, Jenny Fiore, Greg Pearson—and Lee Huntsman, part-time mentor, full-time friend.

I'd also like to hand out three congressional medals of tolerance to my patient family: my wife, Kari, and our sons, Joshua and Noah. We were fortunate to be able to huddle together in one place throughout the viral crisis—and yet a book had to be written in its midst. You three tolerated my need to isolate downstairs, occasionally coming up for air, invariably to be met with crazy cool music, homemade pizza, homemade love. I'm grateful we could assemble in our home for such a long time—one last full recapitulation of the familial rhythms we shared when we were all so much younger. I will treasure it forever.

about the author

DR. JOHN J. MEDINA is a developmental molecular biologist focused on the genes involved in human brain development and the genetics of psychiatric disorders. He has spent most of his professional life as a private research consultant, working in the biotechnology and pharmaceutical industries on research related to mental health. He also consults with non-biotech companies on issues related to management, education, and innovation, including Apple, Boeing, Microsoft, and architecture firms such as NBBJ.

The founding director of two brain research institutes, Medina holds an affiliate faculty appointment at the University of Washington School of Medicine, in its department of bioengineering.

In 2004, Medina was appointed to the rank of affiliate scholar at the National Academy of Engineering. He has been named Outstanding Faculty of the Year at the College of Engineering at the University of Washington; the Merrill Dow Medical Education National Teacher of the Year; and, twice, the Bioengineering Student Association Teacher of the Year. Medina has been a consultant to the

Education Commission of the States and a regular speaker on the relationship between cognitive neuroscience and education.

Medina's books include *Brain Rules: 12 Principles for Surviving and Thriving at Work, Home, and School*; *Brain Rules for Baby: How to Raise a Smart and Happy Child from Zero to Five*; *Brain Rules for Aging Well: 10 principles for staying vital, Happy, and Sharp*; *Attack of the Teenage Brain! Understanding and Supporting the Weird and Wonderful Adolescent Learner*; *The Genetic Inferno*; *The Clock of Ages*; *Depression: How It Happens, How It's Healed*; *What You Need to Know About Alzheimer's*; *The Outer Limits of Life*; *Uncovering the Mystery of AIDS*; and *Of Serotonin, Dopamine, and Antipsychotic Medications*. He also authored and presented *Your Best Brain* with The Teaching Company of Chantilly, Virginia, a twelve-hour Great Courses lecture series describing basic cognitive neuroscience.

Medina has a lifelong fascination with how the mind reacts to and organizes information. As a husband and father of two boys, he has an interest in how the brain sciences might influence the way we teach our children.

In addition to his research, consulting, and teaching, Medina speaks often to public officials, business and medical professionals, school boards, and nonprofit leaders.

index